세상에서 가장 쉬운 과학 수업

쿼크 & 끈

세상에서 가장 쉬운 과학 수업

ⓒ 정완상, 2025

초판 1쇄 인쇄 2025년 12월 5일
초판 1쇄 발행 2025년 12월 15일

지은이 정완상
펴낸이 이성림
펴낸곳 성림북스

책임편집 최윤정
디자인 쏘울기획

출판등록 2014년 9월 3일 제25100-2014-000054호
주소 제주특별자치도 제주시 한경면 고산서3길 135
대표전화 064-772-5762
팩스 064-773-5762
이메일 sunglimonebooks@naver.com

ISBN 979-11-24072-06-6   03400

* 책값은 뒤표지에 있습니다.
* 이 책의 판권은 지은이와 성림북스에 있습니다.
* 이 책의 내용 전부 또는 일부를 재사용하려면 반드시 양측의 서면 동의를 받아야 합니다.

노벨상 수상자들의 **오리지널 논문으로 배우는 과학**

세상에서 가장 쉬운 과학 수업
쿼크소년

정완상 지음

**입자가속기의 발명부터 쿼크의 발견까지
보이지 않는 입자를 예측한 겔만의 위대한 논문 속으로**

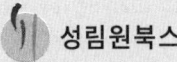

CONTENTS

과학을 처음 공부할 때 이런 책이 있었다면 얼마나 좋았을까 008
천재 과학자들의 오리지널 논문을 이해하게 되길 바라며 011
보이지 않는 입자를 예측한 겔만 _ 윌첵 박사 깜짝 인터뷰 015

첫 번째 만남
입자가속기의 발명 / 021

콕크로프트와 월턴의 가속기 _ 최초의 입자가속기 022
밴더그래프 가속기 _ 고전압 발전기를 발명하다 025
선형가속기 _ 가속기 발전의 역사 027
로런스의 사이클로트론 _ 종이 냅킨의 스케치로부터 032

싱크로트론 _ 입자의 궤도를 나선에서 원형으로　　　　　　　　038

두 번째 만남
군과 대수 / 045

갈루아의 생애 _ 비운의 천재 수학자　　　　　　　　　　　　046
군의 정의 _ 이항연산과 네 가지 조건　　　　　　　　　　　　051
회전군 _ 반시계 방향으로 몇 도 회전시킬까?　　　　　　　　　057
마리우스 솝후스 리 _ 동료들과 함께 학문적 열정을 쏟다　　　064
행렬군과 리군 _ 행렬군이 특별한 관계를 만족할 때　　　　　　067
유니터리군 _ N차원 복소 공간에서　　　　　　　　　　　　　076
리대수 _ 변수가 아주 작은 경우　　　　　　　　　　　　　　081

세 번째 만남
새로운 입자의 발견 / 087

새로운 입자 _ 입자 발견의 역사　　　　　　　　　　　　　　088
아이소스핀의 등장 _ 동일한 입자의 서로 다른 두 상태　　　　095
중간자 _ 핵력을 매개하는 입자가 필요해!　　　　　　　　　　096
타우 입자 _ 전자나 뮤온보다 무거운 경입자를 찾아서　　　　　101
약력을 매개하는 입자 _ 질량이 0이 아닌 새로운 입자　　　　　103
약력을 매개하는 입자의 발견 _ 양성자와 반양성자를 충돌시켜　108

네 번째 만남
기묘한 입자의 출현 / 111

케이온을 발견한 두 과학자 _ 파이온 다음으로 발견된 중간자   112

V입자 _ 성차별을 극복하고   113

케이온 _ 기묘한 입자   123

기묘도 _ 입자의 속성을 나타내는 새로운 수치   126

새로운 중입자 _ 람다, 시그마, 크시, 델타   129

니시지마-나카노-겔만 법칙 _ 세 사람이 발견한 재미있는 공식   133

다섯 번째 만남
쿼크모형 / 137

사카타 모형 _ 페르미-양 모형을 모든 강입자로 확장하다   138

쿼크모형의 창시자 겔만 _ 겔만의 생애와 연구   143

강입자의 정리 _ 짝지어 배열하기   145

겔만의 쿼크모형 _ 제임스 조이스의 소설에서   147

su(3)대수와 쿼크모형 _ 대칭군과 연결 짓다   155

여섯 번째 만남
쿼크모형의 진화 / 161

한무영과 난부의 색깔이 있는 쿼크 _ 빛의 삼원색처럼   162

참 쿼크의 존재를 예언한 과학자들 _ 비요르켄, 일리오풀로스, 마이아니, 이휘소   170
참 쿼크의 필요성 _ 입자가속기의 실험 결과로부터   174
글루온 _ 마치 고무줄처럼   177
참 쿼크의 발견 _ 독립적으로 찾아낸 새로운 중간자   180
보텀 쿼크와 톱 쿼크 _ 이론에서 증거 수집을 거쳐 발견에 이르기까지   184

만남에 덧붙여 / 187

The Production of High Speed Light Ions Without the Use of High Voltages_로런스 논문 영문본   188
Charge Independence for V-particles_나카노-니시지마 논문 영문본   207
On a Composite Model for the New Particles_사카타 논문 영문본   209
Symmetries of Baryons and Mesons_겔만 논문 영문본   212
Discovery of a Narrow Resonance in $e^+ e^-$ Annihilation_릭터 논문 영문본   230
Experimental Observation of a Heavy Particle J_팅 논문 영문본   233
위대한 논문과의 만남을 마무리하며   236
이 책을 위해 참고한 논문들   239
수식에 사용하는 그리스 문자   242
노벨 물리학상 수상자들을 소개합니다   243

과학을 처음 공부할 때 이런 책이 있었다면 얼마나 좋았을까

남순건(경희대학교 이과대학 물리학과 교수 및 전 부총장)

21세기를 20여 년 지낸 이 시점에서 세상은 또 엄청난 변화를 맞이하리라는 생각이 듭니다. 100년 전 찾아왔던 양자역학은 반도체, 레이저 등을 위시하여 나노의 세계를 인간이 이해하도록 하였고, 120년 전 아인슈타인에 의해 밝혀진 시간과 공간의 원리인 상대성이론은 이 광대한 우주가 어떤 모습으로 만들어져 왔고 앞으로 어떻게 진화할 것인가를 알게 해주었습니다. 게다가 우리가 사용하는 모든 에너지의 근원인 태양에너지를 핵융합을 통해 지구상에서 구현하려는 노력도 상대론에서 나오는 그 유명한 질량-에너지 공식이 있기에 조만간 성과가 있을 것이라 기대하게 되었습니다.

앞으로 올 22세기에는 어떤 세상이 펼쳐질지 매우 궁금합니다. 특히 인공지능의 한계가 과연 무엇일지, 또한 생로병사와 관련된 생명의 신비가 밝혀져 인간 사회를 어떻게 바꿀지, 우주에서는 어떤 신비로움이 기다리고 있는지, 우리는 불확실성이 가득한 미래를 향해 달려가고 있습니다. 이러한 불확실한 미래를 들여다보는 유리구슬 역할을 하는 것이 바로 과학적 원리들입니다.

지난 백여 년간 과학에서의 엄청난 발전들은 세상의 원리를 꿰뚫어보았던 과학자들의 통찰을 통해 우리에게 알려졌습니다. 이런 과학 발전을 가능하게 한 영웅들의 생생한 숨결을 직접 느끼려면 그들이 썼던 논문들을 경험해보는 것이 좋습니다. 그런데 어느 순간 일반인과 과학을 배우는 학생들은 물론, 그 분야에서 연구를 하는 과학자들마저 이런 숨결을 직접 경험하지 못하고 이를 소화해서 정리해놓은 교과서나 서적들을 통해서만 접하고 있습니다. 창의적인 생각의 흐름을 직접 접하는 것은 그런 생각을 했던 과학자들의 어깨 위에서 더 멀리 바라보고 새로운 발견을 하고자 하는 사람들에게 매우 중요합니다.

저자인 정완상 교수가 새로운 시도로써 이러한 숨결을 우리에게 전해주려 한다고 하여 그의 30년 지기인 저는 매우 기뻤습니다. 그는 대학원생 때부터 당시 혁명기를 지나면서 폭발적인 발전을 하고 있던 끈 이론을 위시한 이론물리학 분야에서 가장 많은 논문을 썼던 사람입니다. 그리고 그러한 에너지가 일반인들과 과학도들을 위한 그의 수많은 서적을 통해 이미 잘 알려져 있습니다. 저자는 이번에 아주 새로운 시도를 하고 있고 이는 어쩌면 우리에게 꼭 필요했던 것일 수 있습니다. 대화체로 과학의 역사와 배경을 매우 재미있게 설명하고, 그 배경 뒤에 나왔던 과학 영웅들의 오리지널 논문들을 풀어간 것입니다. 과학사를 들려주는 책들은 많이 있으나 이처럼 일반인과 과학도의 입장에서 질문하고 이해하는 생각의 흐름을 따라 설명한 책은 없습니다. 게다가 이런 준비를 마친 후에 아인슈타인 같은 영웅들의 논문을 원래의 방식과 표기를 통해 설명하는 부분은 오랫동안 과학을 연구해온 과학자에게도 도움을 줍니다.

    이 책을 읽는 독자들은 복 받은 분들일 것이 분명합니다. 제가 과학을 처음 공부할 때 이런 책이 있었다면 얼마나 좋았을까 하는 생각이 듭니다. 정완상 교수는 이제 새로운 형태의 시리즈를 시작하고 있습니다. 독보적인 필력과 독자에게 다가가는 그의 친밀성이 이 시리즈를 통해 재미있고 유익한 과학으로 전해지길 바랍니다. 그리하여 과학을 멀리하는 21세기의 한국인들에게 과학에 대한 붐이 일기를 기대합니다. 22세기를 준비해야 하는 우리에게는 이런 붐이 꼭 있어야 하기 때문입니다.

천재 과학자들의 오리지널 논문을
이해하게 되길 바라며

　사람들은 과학 특히 물리학 하면 너무 어렵다고 생각하지요. 제가 외국인들을 만나서 얘기할 때마다 신선하게 느끼는 점이 있습니다. 그들은 고등학교까지 과학을 너무 재미있게 배웠다고 하더군요. 그래서인지 과학에 대해 상당한 지식을 가진 사람들이 많았습니다. 그 덕분에 노벨 과학상도 많이 나오는 게 아닐까 생각해요. 우리나라는 노벨 과학상 수상자가 한 명도 없습니다. 이제 청소년과 일반 독자의 과학 수준을 높여 노벨 과학상 수상자가 매년 나오는 나라가 되게 하고 싶다는 게 제 소망입니다.

　그동안 양자역학과 상대성이론에 관한 책은 전 세계적으로 헤아릴 수 없을 정도로 많이 나왔고 앞으로도 계속 나오겠지요. 대부분의 책은 수식을 피하고 관련된 역사 이야기들 중심으로 쓰여 있어요. 제가 보기에는 독자를 고려하여 수식을 너무 배제하는 것 같았습니다. 이제는 독자들의 수준도 많이 높아졌으니 수식을 피하지 말고 천재 과학자들의 오리지널 논문을 이해하길 바랐습니다. 그래서 앞으로 도래할 양자(量子, quantum)와 상대성 우주의 시대를 멋지게 맞이하도록 도우리라는 생각에서 이 기획을 하게 된 것입니다.

원고를 쓰기 위해 논문을 읽고 또 읽으면서 어떻게 이 어려운 논문을 독자들에게 알기 쉽게 설명할까 고민했습니다. 여기서 제가 설정한 독자는 고등학교 정도의 수식을 이해하는 청소년과 일반 독자입니다. 물론 이 시리즈의 논문에 그 수준을 넘어서는 내용도 나오지만 고등학교 수학만 알면 이해할 수 있도록 설명했습니다. 이 책을 읽으며 천재 과학자들의 오리지널 논문을 얼마나 이해할지는 독자들에 따라 다를 거라 생각합니다. 책을 다 읽고 100% 혹은 70%를 이해하거나 30% 미만으로 이해하는 독자도 있을 것입니다. 제 생각으로는 이 책의 30% 이상 이해한다면 그 사람은 대단하다고 봅니다.

이 책에서는 쿼크 이론이 등장하는 과정을 여섯 편의 논문(로런스, 나카노-니시지마, 사카타, 겔만, 릭터, 팅)과 함께 다루었습니다.

20세기 중반 물리학자들은 원자와 원자핵을 넘어서 그 안의 더 작은 세계로 눈을 돌렸습니다. 뮤온과 타우 입자 같은 경입자(lepton)가 발견되었고, 이들을 따라다니는 뮤온 뉴트리노, 타우 뉴트리노도 모습을 드러냈습니다. 이 책에서는 경입자 발견의 역사도 소개하고 있습니다. 또한 양성자와 중성자 같은 중입자, 파이온과 케이온 같은 중간자(meson) 발견도 이야기했습니다. 그중 케이온의 등장과 함께 도입된 기묘도(strangeness)라는 양자수를 다루었고, 여기에 초전하(hypercharge), 아이소스핀(isospin)을 리군과 리대수로 묘사했습니다.

더불어 1964년 등장하는 겔만의 쿼크모형과 이 모형이 예측한 오메가 입자의 발견을 다룹니다. 하나의 입자로 여겨졌던 양성자와 중성자가 더 작은 기본 입자인 쿼크로 구성되어 있다는 것을 독자 여러분에게 소개합니다. 겔만이 도입한 u, d, s 쿼크와 그 이후에 예언된 참 쿼크, 보텀 쿼크, 톱 쿼크의 발견도 곁들였습니다. 이 책은 그 작은 입자들의 세계 속에서 인류가 어떻게 숨겨진 질서를 발견해 왔는지를 따라가는 이야기입니다.

〈노벨상 수상자들의 오리지널 논문으로 배우는 과학〉 시리즈는 많은 이에게 도움을 줄 수 있다고 생각합니다. 과학자가 꿈인 학생과 그의 부모, 어릴 때부터 수학과 과학을 사랑했던 어른, 양자역학과 상대성이론을 좀 더 알고 싶은 사람, 아이들에게 위대한 논문을 소개하려는 과학 선생님, 반도체나 양자암호 시스템, 우주 항공 계통 등의 일에 종사하는 직장인, 〈인터스텔라〉를 능가하는 SF 영화를 만들고 싶어 하는 영화 제작자나 웹툰 작가 등 많은 사람들에게 이 시리즈를 추천합니다.

끝으로 용기를 내서 이 시리즈의 출간을 결정한 성림원북스의 이성림 사장과 직원들에게 감사를 드립니다. 시리즈 초안이 나왔을 때, 수식이 많아 출판사들이 꺼릴 것 같다는 생각이 들었습니다. 몇 군데에 출판을 의뢰한 후 거절당하면 블로그에 올릴 생각으로 글을 써 내려갔습니다. 놀랍게도 첫 번째로 이 원고의 이야기를 나눈 성림원북스에서 출간을 결정해 주어서 책이 나올 수 있게 되었습니다. 원고를 쓰는 데 필요한 프랑스 논문의 번역을 도와준 아내에게도 고마움을 전합니다. 그리고 이 책을 쓸 수 있도록 멋진 논문을 만든 고 겔만 박사님에게도 감사를 드립니다.

진주에서 정완상 교수

보이지 않는 입자를 예측한 겔만
_ 윌첵 박사 깜짝 인터뷰

천재적인 직관으로 입자물리학을 체계화하다

기자   오늘은 1964년 발표된 겔만의 쿼크모형 논문에 관해 윌첵 박사와 인터뷰를 진행하겠습니다. 윌첵 박사는 점근적 자유 모형으로 2004년 노벨 물리학상을 수상한 분이지요. 윌첵 박사님, 나와 주셔서 감사합니다.

윌첵   제가 제일 존경하는 과학자인 겔만의 논문에 관한 내용이라 만사를 제치고 달려왔습니다.

기자   겔만은 어떤 인물인가요?

윌첵   겔만에 대해 단순히 '똑똑하다'는 말로는 부족합니다. 그는 복잡한 것들 속에서 패턴을 찾아내는 천재적인 직관을 가진 사람이었습니다. 입자물리학이 혼란스럽고 조각난 퍼즐 같던 시절, 그는 수많은 입자를 '맛'과 '향기'로 나눠 체계화하기 시작했죠. 마치 정리되지 않은 동물들을 분류하던 고대 자연사학자처럼 말이에요.

기자   맛과 향기요? 물리학 이야기 맞나요?

윌첵   그렇습니다. 겔만은 입자들의 정체를 설명하기 위해 아주 독창적인 언어와 개념을 사용했어요.

기자   흥미롭군요.

자연의 아름다움을 가장 잘 이해한 물리학자

기자   겔만의 1964년 논문에는 어떤 내용이 들어 있나요?

윌첵   그가 쓴 논문은 짧지만, 입자물리학의 판도를 완전히 바꾼 혁명적인 선언문이었습니다. 거기에서는 우리가 알고 있는 양성자, 중성자 같은 입자가 사실은 더 기본적인 입자인 쿼크 3개로 구성되어 있다는 이론을 제안합니다. 당시엔 실험으로 관측되지도 않은 입자들을, 단지 수학적 대칭성과 패턴을 통해 예측한 것이죠.

기자   쿼크라는 단어도 참 독특하던데요?

윌첵   겔만은 제임스 조이스의 소설 《피네간의 경야》에서 "Three quarks for Muster Mark!"라는 구절을 보고 그 단어를 따와서 썼다고 해요. '쿼크'라는 이름은 말장난처럼 들리지만, 사실 그 안엔 겔만 특유의 문학적 감성, 직관, 유머, 그리고 물리학적 깊이가 모두 담겨 있습니다.

기자   물리학계에서는 어떻게 받아들였나요?

윌첵   처음에는 다들 "너무 멀리 나간 상상"이라고 생각했죠. "관측되지도 않은 입자를 믿으라고?"라는 반응도 나왔고요. 하지만 시간이 지나면서 겔만의 이론은 점점 더 많은 입자 간 관계를 설명해 주었어요. 결국 입자물리학의 주류 이론인 표준모형(Standard Model)의 기초가 되었습니다. 겔만이 아니었다면 오늘날의 입자물리학도 없었을 겁니다.

기자   겔만은 과학자이자 예술가였군요.

윌첵　　정확한 표현입니다. 그는 자연의 대칭성과 아름다움을 가장 잘 이해한 물리학자 중 한 명이었어요. 그의 세계관은 수식이 아닌 '조화'로 설명되었죠.

기자　　그때는 쿼크가 관측되지도 않았잖아요?

윌첵　　맞습니다. 쿼크는 당시 실험으로 직접 본 사람이 아무도 없었어요. 그런데 겔만은 입자들 사이의 복잡한 관계, 즉 대칭성과 분류 체계를 분석하면서 "이 복잡한 퍼즐을 풀 수 있는 더 간단한 조각들이 필요하다"는 걸 꿰뚫어 보았죠. 그는 $SU(3)$군이라는 수학 개념을 이용해, 기존 입자들을 3개의 쿼크(u, d, s) 조합으로 설명했어요.

기자　　수학으로 입자를 예측했다는 말인가요?

윌첵　　그래요. 겔만은 관찰된 데이터의 대칭성과 규칙성만 보고, 눈에 보이지 않는 입자를 예측한 겁니다. 마치 주기율표의 빈칸을 보고 새로운 원소를 예측한 멘델레예프처럼, 과학사에서 손꼽히는 '이성의 직관'이었죠.

기자　　왜 사람들이 처음엔 반신반의했을까요?

윌첵　　그도 그럴 것이 겔만이 제안한 쿼크는 정수 전하가 아닌 1/3이나 2/3 전하를 가졌거든요. 당시 물리학자들은 "자연에 그런 전하가 존재할 리 없다"고 생각했죠. 게다가 쿼크는 자유로운 상태로는 결코 관측되지 않는다는 점도 받아들이기 어려웠고요.

기자　　결국은 이론이 맞았다는 거네요.

윌첵　　시간이 흐르면서 쿼크가 존재한다는 증거가 나왔어요. 오늘날 말하는 표준모형에서 쿼크는 핵심 구성 요소가 되었습니다.

기자　　더 자세히 알아보고 싶네요.

입자물리학의 세계관을 완전히 바꾸다

기자　　겔만의 1964년 논문은 어떤 변화를 가져왔나요?

윌첵　　그의 논문은 한 편의 이론 제안서로 그친 게 아니라, 입자물리학의 세계관을 완전히 바꾼 철학적 전환점이었습니다. 그 전까지 과학자들은 양성자, 중성자, 파이온, 시그마 같은 입자들을 각기 다른 기본 입자라고 생각했어요. 겔만은 그 수많은 입자가 사실은 소수의 쿼크라는 기본 벽돌로 지어진 구조물이라고 제안했습니다. 그건 곧 "우주를 구성하는 단위가 우리가 생각한 것보다 훨씬 더 단순하고 아름답다"는 선언이었죠.

기자　　물리학자들의 시야가 달라졌다는 말인가요?

윌첵　　정확합니다. 겔만의 논문은 '현상 중심 물리학'에서 '구조 중심 물리학'으로 전환을 이끌었어요. 전에는 입자가 보이면 그걸 이름 붙이고 분류했습니다. 하지만 겔만 이후엔 "이 입자는 어떤 쿼크 조합으로 이루어졌나?", "어떤 대칭 원리에 의해 설명되나?"를 묻기 시작했죠.

기자　　단순한 이론이 아니라 '생각하는 틀'이 달라진 거군요.

윌첵　　맞아요. 겔만은 입자들의 질서와 분류, 그 뒤에 숨겨진 조화를 보여주었습니다. 그게 나중에 표준모형의 뼈대가 되었고요. 우리는 더 이상 입자를 '발견'하는 것이 아니라, 수학적 구조 속에서 '예측'하게 됐습니다.

기자   겔만의 논문은 쿼크를 제안했지만, 실험으로 바로 증명된 건 아니잖아요?

윌첵   그게 오히려 겔만의 위대함을 더 드러내는 부분입니다. 그는 보이지 않는 것을 상상했고, 그것이 실제로 존재한다는 믿음이 있었습니다. 쿼크는 자유롭게 존재하지 않고, 항상 얽혀 있어서 관측되지 않는 점까지 간파했죠. 그리고 수십 년 뒤, 우리가 딥인엘라스틱 산란 실험과 중입자 붕괴 데이터를 통해 그의 예언이 맞았다는 걸 하나씩 확인하게 된 겁니다.

기자   결국 겔만의 논문은 '어떻게 보느냐'를 바꾼 거네요.

윌첵   그렇죠. 그는 복잡해 보이는 우주의 겉모습 뒤에 숨어 있는 단순함, 대칭성, 조화의 원리를 보여주었어요. 그는 물리학자이면서 동시에 형이상학자, 그리고 예술가였습니다. 그가 쓴 수식은 단지 계산이 아니라, 세계에 대한 하나의 시(詩)였죠.

기자   그렇군요. 지금까지 겔만의 쿼크모형 논문에 대해 윌첵 박사의 이야기를 들어 보았습니다.

첫 번째 만남

입자가속기의 발명

콕크로프트와 월턴의 가속기 _최초의 입자가속기

정교수 쿼크 발견을 이야기하려면 먼저 입자가속기의 역사를 살펴봐야 해.

물리군 입자를 왜 가속시키죠?

정교수 입자를 가속시켜서 충돌을 통해 새로운 입자를 발견하기 위해서야. 그러려면 입자에 힘을 작용해야 하는데 초기의 가속기는 전기력으로 입자를 가속시키는 장치였어. 큰 전기력을 작용하기 위해서는 높은 전압을 발생시키는 장치를 만들어야 하지. 1920년대 말에 여러 과학자가 전기력을 이용한 가속기를 발명했어. 이제 콕크로프트-월턴 가속기를 알아볼 거야. 먼저 콕크로프트를 소개할게.

콕크로프트(John Douglas Cockcroft, 1897~1967, 1951년 노벨 물리학상 수상)

콕크로프트는 1897년 영국 요크셔의 토드모든에서 태어났다. 그는 토드모든에서 초중고를 마친 후 맨체스터 빅토리아 대학에서 수학을 공부했다.

1914년 8월에 제1차 세계대전이 발발했다. 콕크로프트는 이듬해 11월 24일, 영국 육군에 입대했다. 전쟁이 끝난 뒤에는 맨체스터 빅토리아 대학으로 돌아가지 않고, 맨체스터 시립 공과대학에서 전기 공학을 공부해 1920년에 졸업했다. 그는 1922년 〈교류에 대한 고주파 분석〉 논문으로 석사 학위를 받았다. 그 후 러더퍼드의 지도로 1925년에 박사 학위를 받았다.

이번에는 월턴에 대해 알아보자.

월턴(Ernest Thomas Sinton Walton, 1903~1995, 1951년 노벨 물리학상 수상)

월턴은 아일랜드의 던가번에서 태어났다. 그는 1922년 더블린 트리니티 칼리지에서 수학과 과학을 공부하고 1927년에 석사 학위를 받았다. 그 후 영국 케임브리지 대학 캐번디시 연구소의 연구원이 되었다. 당시 이 연구소의 소장은 러더퍼드였다.

물리군 두 사람은 캐번디시 연구소에서 만났군요.

정교수 맞아. 1919년에 러더퍼드는 붕괴하는 라듐 원자에서 방출되는 알파 입자로 질소 원자를 분해하는 데 성공했어. 이 연구를 위해 전기를 띤 입자를 가속시키는 장치를 만들어야 했지. 1928년 러더퍼드는 이 문제를 콕크로프트와 월턴에게 맡겼고, 두 사람은 콕크로프트-월턴 가속기라고 부르는 최초의 가속기를 만드는 데 성공했다네.

콕크로프트-월턴 가속기
(출처: Geni/Wikimedia Commons)

1932년 4월 14일, 콕크로프트와 월턴은 양성자를 가속시켜 리튬과 충돌하게 하여 리튬이 헬륨으로 변하는 것을 알아냈어. 이것은 최초의 인공 핵변환 실험이었고, 이 업적으로 두 사람은 노벨 물리학상을 받게 되지.

밴더그래프 가속기 _ 고전압 발전기를 발명하다

정교수 전기력을 이용해 전기를 띤 입자를 가속시키려면 높은 전압을 발생시키는 장치인 고전압 발전기가 필요해. 이번에는 고전압 발전기를 발명한 밴더그래프에 대해 이야기해 볼게.

밴더그래프(Robert Jemison Van de Graaff, 1901~1967)

밴더그래프는 미국 앨라배마주 터스컬루사에서 태어났다. 그는 앨라배마 대학에서 수학 학사(1922)와 기계공학 석사(1923) 학위를 취득했다. 앨라배마 전력 회사에서 1년 동안 일한 후, 밴더그래프는 1925년 프랑스 소르본 대학에서 마리 퀴리의 강의에 참석했다. 1926년 그는 옥스퍼드 대학에서 두 번째 학사 학위를, 1928년에는 같은 대학에서 존 실리 타운센드의 지도로 박사 학위를 받았다.

1929년 밴더그래프는 8만 볼트의 고전압을 발생시키는 발전기를 개발했다. 1933년에는 700만 볼트의 고전압을 발생시키는 데 성공했다.

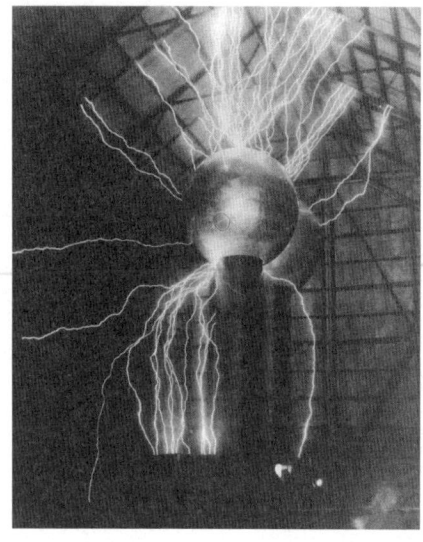

밴더그래프 발전기

밴더그래프는 1929년부터 1931년까지 프린스턴 대학 교수로 지냈고, 이후 매사추세츠 공과대학으로 자리를 옮겼다.

제2차 세계대전 당시 밴더그래프는 고전압 방사선 촬영 프로젝트의 책임자였다. 전쟁이 끝난 뒤에는 트럼프(John G. Trump)와 함께

HVEC(High Voltage Engineering Corporation)라는 회사를 공동 설립했다. 1950년대에는 고전압 직류를 생산하는 절연 코어 변압기를 발명했다.

선형가속기 _ 가속기 발전의 역사

정교수   이번에는 선형가속기를 발명한 과학자들을 소개할 거야. 선형가속기 아이디어는 1924년 스웨덴의 구스타브 이징이 처음 발표했어. 하지만 그는 가속기를 만드는 데는 실패했지.

이징(Gustaf Ising, 1883~1960,
사진 출처: Ryanicus Girraficus/Wikimedia Commons)

물리군   어떤 원리인가요?
정교수   고전압에 의해 전자를 가속시키는 거야. 여러 개의 튜브를 이용해 가속된 전자에 다시 고전압을 작용해서 전자의 가속도를 극대화하는 장치라네.

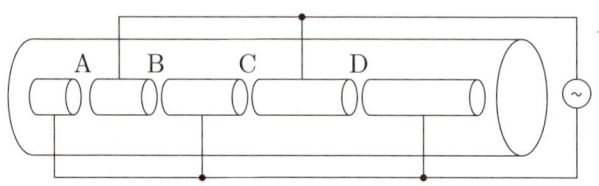

물리군 선형가속기를 처음 만든 사람은 누구예요?

정교수 노르웨이의 비데뢰야.

비데뢰(Rolf Widerøe, 1902~1996,
사진 출처: ETH-Bibliothek/Wikimedia Commons)

비데뢰는 1902년 노르웨이의 크리스티아니아(현재 오슬로)에서 태어났다. 그는 오슬로의 할링 스쿨을 졸업하고, 독일 카를스루에 대학에서 전기공학을 공부했다.

1924년 비데뢰는 노르웨이로 돌아와 노르웨이 국영 철도의 기관차 시설에서 일했다. 1925년 그는 아헨 공과대학에서 공부했으며, 1928년에 스웨덴 과학자 구스타브 이징의 이론을 토대로 최초의 선형가속기를 만드는 데 성공했다.

여기서 잠깐 에너지 단위인 전자볼트(eV)에 대해 알아보자. 1eV는 1볼트의 전압이 걸려 전자가 가속될 때 전자의 에너지를 말한다. 즉, 전자의 전하량과 전압의 곱이다. 그러니까 1eV를 J로 바꾸면

$1eV = 1.6 \times 10^{-19} J$

이다. 1eV의 1000배를 1keV, 1keV의 1000배를 1MeV, 1MeV의 1000배를 1GeV, 1GeV의 1000배를 1TeV라고 한다.

1keV(킬로전자볼트) = 1000eV
1MeV(메가전자볼트) = 1000keV
1GeV(기가전자볼트) = 1000MeV
1TeV(테라전자볼트) = 1000GeV

비데뢰는 선형가속기를 이용해 나트륨과 칼륨 이온을 5만 전자볼트(50keV)의 에너지로 가속시키는 데 성공했다. 선형가속기는 당시 쓰이던 정전기 입자가속기(콕크로프트-월턴 가속기 및 밴더그래프 가속기)보다 더 높은 입자 에너지를 생성할 수 있었다.

입자가속기는 계속 발전해 1947년에 미국의 앨버레즈(Luis

Alvarez)는 양성자를 31.5MeV의 에너지로 가속시키는 데 성공했다.

앨버레즈의 선형가속기
(출처: Justin Clements/Wikimedia Commons)

선형가속기는 암 치료에 쓰였다. 선형가속기 기반 방사선 치료는 1953년 영국 런던의 해머스미스 병원에 처음 등장했다. 이때 사용된 것은 8메가전자볼트 선형가속기였다. 얼마 후인 1954년, 미국 스탠퍼드에 6메가전자볼트 선형가속기가 설치되어 1956년에 치료를 시작했다.

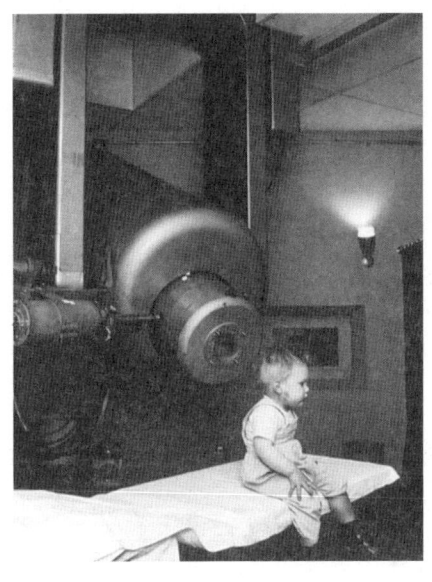

1957년 미국에서 선형가속기 방사선 요법으로 망막모세포종을 치료한 최초의 환자 고든 아이작스(Gordon Isaacs)

입자물리학 실험을 위한 선형가속기는 스탠퍼드 대학에 처음 만들어졌다. 이 가속기의 이름은 Stanford Linear Accelerator Center로 줄여서 SLAC라고 부르는데, 초전도자석을 이용한 선형가속기이다. 1966년에 건설된 SLAC는 길이가 3.2킬로미터인 세계 최대의 선형가속기로, 전자를 50GeV의 에너지로 가속시킬 수 있다.

SLAC의 항공 사진

SLAC 터널

로런스의 사이클로트론 _ 종이 냅킨의 스케치로부터

정교수 이제 입자가속기로 노벨 물리학상을 받은 로런스의 이야기를 해볼게.

로런스(Ernest Orlando Lawrence, 1901~1958, 1939년 노벨 물리학상 수상)

로런스는 1901년 미국 사우스다코타주 캔턴에서 태어났다. 그의 아버지는 노르웨이 이민자였고 캔턴의 고등학교 교사였다.

로런스는 캔턴과 피어의 공립학교에 다녔다. 그는 미네소타주 노스필드에 있는 세인트 올라프 대학에 입학했으나, 1년 뒤 버밀리언의 사우스다코타 대학으로 편입했다. 1922년에 그는 화학 학사 학위를 받았다. 1923년에는 미네소타 대학에서 윌리엄 프랜시스 그레이 스완(William Francis Gray Swann)의 지도하에 물리학 석사(M.A.) 학위를 마쳤다. 석사 논문을 위해 로런스는 자기장으로 타원체를 회전시키는 실험 장치를 만들었다.

스완을 따라 시카고 대학에 진학한 로런스는 이후 코네티컷주 뉴헤이븐에 있는 예일 대학으로 옮겼다. 그곳에서 그는 1925년에 칼륨 증기의 광전효과 연구로 물리학 박사 학위를 취득하고, 예일 대학 연구원이 되었다.

버지니아 대학의 제시 빔스(Jesse Beams)와 함께 로런스는 광전효과 연구를 계속했다. 그들은 광자가 광전 표면에 부딪힌 후 2×10^{-9}초 이내에 광전효과가 발생하는 것을 알아냈다.

로런스는 1928년부터 캘리포니아 대학의 물리학 부교수로 2년을 지낸 다음 최연소 정교수가 되었다. 그는 자신의 실험실에서 고에너지 양성자를 탄소-13 원소에 발사하여 질소-13 동위원소를 발견했다. 이는 프레데리크(Frédéric)와 이렌 졸리오퀴리(Irène Joliot-Curie)가 1934년에 발표한 인공 방사능 연구를 바탕으로 한 것이었다.

로런스에게 노벨 물리학상을 안겨준 것은 사이클로트론이라는 입자가속기이다. 이 발상은 종이 냅킨에 스케치하는 데서 시작했다.

1929년 어느 날 저녁, 도서관에 앉아 있던 로런스는 비데뢰의 연구 결과를 보았다. 그리고 전기장에 의해 전자가 가속되듯 자기장으로도 전자를 가속시킬 수 있다는 것을 떠올렸다.

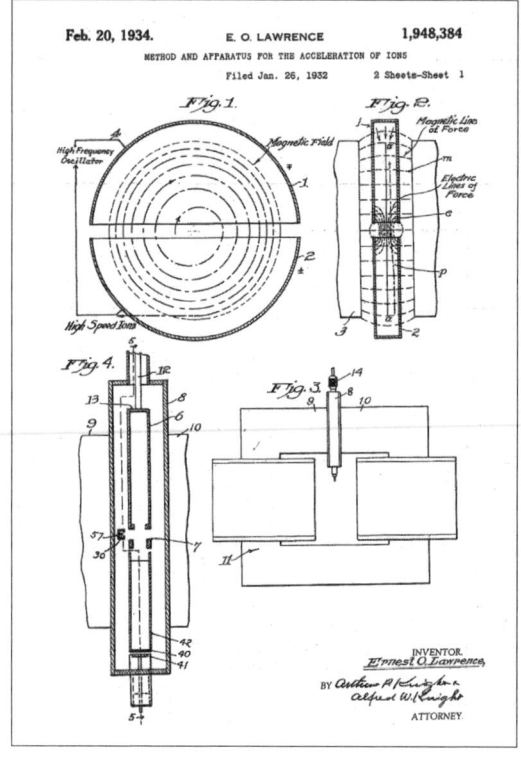

1934년 특허로 제출된 사이클로트론 원리

사이클로트론은 높은 진동수의 교류전압으로 전기를 띤 입자를 가속시킨다. 교류전압은 진공실 내부에서 D전극으로 불리는 두 개의 속이 빈 D 모양의 금속 전극 사이에 흐른다. 입자들의 움직임을 위해

D전극은 사이에 원통형의 공간을 만들면서 좁은 간격을 두고 마주본다. 입자들은 이 공간의 중심에 투입된다.

이때 전극판과 수직으로 일정한 자기장을 걸어주면 입자들은 자기력을 받아 원형으로 휜다. 만약 입자의 속도가 일정하다면 원형의 경로로 움직인다. 하지만 수천 볼트의 교류전압이 D전극 안에 걸려 있고, 그 주파수는 입자들이 한 번의 전압 주기 동안 하나의 회로를 만들도록 설정되어 있다.

입자들이 다른 D전극을 통과한 후 무선주파수의 방향이 바뀐다. 따라서 입자들이 D전극 사이의 간격을 통과할 때마다 전기장은 그것들을 가속시키기에 알맞은 방향으로 설정된다. 이러한 전기력에 의해 증가한 속도로 입자들은 더 큰 반지름의 원형 경로를 따라 움직인다. 즉, 입자들은 중심에서 가장자리로 향하는 나선의 경로를 따라 움직인다.

D전극의 가장자리에 도착했을 때 입자들은 D전극 사이의 작은 간격을 통해 빠져나온다. 그리고 빠져나온 지점에 있는 목표물과 충돌한 후 사이클로트론을 떠난다. 충돌에 의한 핵반응은 사이클로트론을 나와서 분석을 위한 장치 안으로 유도되는 2차 입자를 생성한다.

이때 입자의 마지막 에너지 E를 구해보자. 입자의 질량을 m, 전하

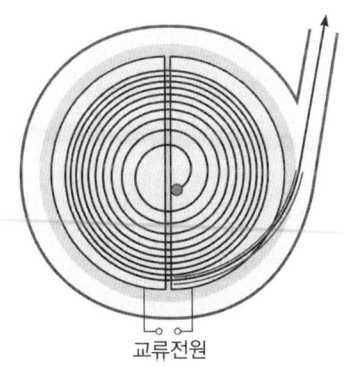

사이클로트론 스케치

량을 q, 입자의 속력을 v, 사이클로트론에 작용한 일정한 자기장의 크기를 B라고 하면 입자가 받는 자기력은

$$(\text{자기력}) = qvB$$

가 된다. 사이클로트론의 반지름을 R이라고 하면 입자가 사이클로트론 속에서 속력 v로 원운동을 할 때 받는 구심력은

$$(\text{구심력}) = \frac{mv^2}{R}$$

이다. 이 두 힘이 평형을 이루므로

$$qvB = \frac{mv^2}{R}$$

에서

$$v = \frac{qBR}{m}$$

임을 알 수 있다. 그러므로 입자의 운동에너지는

$$(\text{운동에너지}) = \frac{1}{2}mv^2 = \frac{q^2}{2m}B^2R^2$$

이다. 즉, 사이클로트론의 반지름이 클수록, 자기장이 셀수록 입자의 운동에너지는 커진다.

로런스는 더 크고 새로운 기계를 만들고자 했다. 1932년 초에 그는 27인치(69센티미터) 사이클로트론을 설계했다. 그는 팰로앨토의

폐차장에서 80톤의 거대한 녹슨 자석을 발견했다. 이는 제1차 세계대전 중 대서양 횡단 무선 링크에 전력을 공급하기 위해 제작된 것이었다. 로런스는 이를 이용해 27인치(69센티미터) 사이클로트론을 만들었다.

그 후 그는 1939년에 60인치(1.52미터) 사이클로트론을 제작했다. 이를 통해 방사성 동위원소들이 발견되었다.

60인치 사이클로트론. 왼쪽에서 세 번째가 로런스(출처: Science Museum London/Wikimedia Commons)

로런스는 의학 연구에 사이클로트론을 쓰도록 권장했다. 핵물리학보다 의료 목적, 특히 암 치료를 위해 돈을 모으는 것이 더 쉬웠기 때문이다. 1938년 캘리포니아 대학에서는 백혈병에 걸린 쥐를 대상으로 한 실험에서 37인치 사이클로트론에서 생성된 인-32를 사용했다.

1939년 11월, 로런스는 '사이클로트론의 발명 및 개발, 특히 인공 방사성 원소와 관련하여 얻은 결과'로 노벨 물리학상을 수상했다.

싱크로트론 _ 입자의 궤도를 나선에서 원형으로

정교수   이번에는 사이클로트론을 개량해 싱크로트론이라는 새로운 입자가속기를 발명한 소련의 벡슬레르에 대해 알아볼게.

벡슬레르(Vladimir Iosifovich Veksler, 1907~1966)

벡슬레르는 1907년 러시아의 지토미르(현재 우크라이나의 도시)에서 태어났다. 그의 가족은 1915년 지토미르에서 모스크바로 이사했다. 1931년에 벡슬레르는 모스크바 전력 공학 연구소를 졸업했다. 1936년부터는 레베데프 물리학 연구소에서 일했으며, 입자검출기 개발과 우주선(cosmic rays) 연구에 참여했다. 벡슬레르는 파미르고원과 엘브루스산을 여러 번 탐험했는데, 그가 속한 탐험대는 우주선의 구성 요소를 집중적으로 연구했다. 1944년부터 그는 가속기 물리학 분야에서 일했으며, 싱크로트론을 개발하는 데 성공했다.

1956년 벡슬레르는 두브나에 있는 합동 원자핵 연구소(JINR)에 고에너지 연구소를 설립하고 초대 소장이 되었다. 이곳에서 당시 세계에서 가장 큰 원형 양성자 가속기를 통합한 싱크로파소트론(Synchrophasotron)을 건설했다. 이는 프로트비노의 가속기와 함께 세워졌다.

싱크로파소트론을 둘러보는 방문자들(출처: Eto shorcy/ Wikimedia Commons)

물리군 싱크로트론은 사이클로트론과 어떻게 다른가요?

정교수 사이클로트론은 입자가 나선 궤도를 그리지만 싱크로트론은 입자가 원형 궤도를 그리지. 이러한 원형 궤도를 유지하기 위해 싱크로트론에서는 자기장의 세기가 시간에 따라 변해. 싱크로트론이 발명된 후에는 입자 실험에서 사이클로트론은 거의 쓰이지 않고 거대한 싱크로트론이 속속들이 개발되었어.

코스모트론 그림

1953년 미국의 브룩헤이븐 국립연구소에 최초의 싱크로트론인 코스모트론이 건설되었다. 이 가속기는 양성자를 3.3GeV까지 가속시키는 데 성공했다. 또한 그동안 우주선에서만 볼 수 있었던 많은 중간자를 관찰했다. 코스모트론의 지름은 75피트(22.9미터)이고, 각각의 무게가 6톤이며 최대 1.5테슬라의 자기장을 생성하는 288개의 자석으로 구성되어 있다.

브룩헤이븐 국립연구소의 코스모트론용 주입 시스템

1954년 미국 로런스 버클리 국립연구소에는 베바트론이라는 이름의 양성자 싱크로트론이 건설되었다. 이듬해인 1955년 그곳에서 반양성자가 발견되었다. 이것으로 1959년 에밀리오 세그레와 오언 체임벌린이 노벨 물리학상을 수상했다.

베바트론

1961년 오스트리아의 물리학자 브루노 투셰크(Bruno Touschek)가 이끈 이탈리아 물리학자 그룹은 ADA(Anello Di Accumulazione) 충돌기라는 싱크로트론을 건설했다. 당시 미국 물리학자들은 고정된 대상에 빔을 쏘는 대신 두 개의 입자 빔을 정면으로 충돌시키는 데 관심이 있었다. ADA는 입자(전자) 빔 중 하나를 반입자 빔(양전자)으로 교체해 최초로 전자-양전자 충돌 실험에 성공했다.

싱크로트론이 내는 에너지는 가속기 원형 링의 반지름을 크게 함으로써 점점 더 증가시킬 수 있었다. 미국에서는 1968년에 일리노이주 바타비아 동쪽의 페르미 국립가속기 연구소(페르미 랩)에 메인 링의 둘레 길이가 6.28킬로미터인 테바트론을 짓기 시작했다. 테바트론은 1973년에서 1979년 사이에 연구 개발 단계에 있었고 메인 링에서의 가속도는 계속 향상되었다.

페르미 국립가속기 연구소의 테바트론

1972년 1월 22일 20GeV, 2월 4일 53GeV, 2월 11일 100GeV, 3월 1일 200GeV로 양성자를 가속시키는 데 성공했다. 1976년 5월 14일에는 양성자를 500GeV까지 가속시켰고, 1984년 2월 16일 800GeV,

1986년 10월 21일 900GeV로 가속도를 높였다. 1986년 11월 30일에는 1.8TeV로 가속시킨 상태에서 양성자-반양성자 충돌을 발생시켰다.

테바트론 터널

물리군 테바트론이 세계에서 가장 큰 에너지를 만드는 입자가속기인가요?

정교수 테바트론은 2011년까지만 운영했어. CERN에 LHC가 만들어지기 전까지는 최대 에너지의 입자가속기였지. 현재 세계에서 제일 크고 에너지가 가장 높은 입자가속기는 프랑스-스위스 국경에 있는 LHC(Large Hadron Collider)야.

LHC는 1998년에서 2008년 사이에 유럽 입자물리 연구소(CERN)가 1만여 명의 과학자와 100개 이상의 국가에 걸쳐 수백 개 대학 및 실험실과 협력하여 건설했다. LHC는 제네바 근처의 프랑스-스위스 국경 아래 둘레가 27킬로미터(17마일), 깊이가 175미터(574피트)인 터널에 있다.

LHC에 있는 입자검출기(출처: SimonWaldherr/Wikimedia Commons)

LHC는 2010년에 3.5TeV의 에너지를 거쳐, 2015년 13.0TeV의 에너지에 도달했다.

두 번째 만남

군과 대수

갈루아의 생애 _ 비운의 천재 수학자

정교수 이번에는 군의 창시자이며, 결투로 요절한 수학 천재 갈루아를 소개할게.

갈루아(Évariste Galois, 1811~1832)

갈루아는 1811년 10월 25일에 프랑스 파리 교외의 부르라렌에서 태어났다. 그의 아버지는 학교 교장이었고 공화주의자로 부르라렌 자유당의 당수였다. 루이 18세가 왕위에 복위된 1814년에는 부르라렌의 시장이 되었다. 갈루아의 어머니는 법률가의 딸로 라틴어 고전 문학을 읽을 수 있었다. 갈루아는 12세까지 어머니에게 교육을 받았다. 그는 10세 때에 프랑스 랭스에 있는 학교에 입학하기를 바랐으나, 어머니는 아직 집에서 배우는 것이 좋겠다고 판단했다.

1823년 갈루아는 리세 루이르그랑에 입학했다. 그해에 정치적인 문제로 100여 명이나 되는 학생들이 퇴학당하는 일이 있었지만, 갈루아는 별다른 문제없이 학교생활을 했다. 2학년 때에는 라틴어 과목에서 우수상을 수상하기도 했다. 그러나 14세 무렵에는 라틴어에 흥미를 잃고 수학에만 몰두했다. 그는 르장드르의 《기하학 기초》를 처음 읽자마자 내용을 숙지하며, 마치 소설처럼 읽었다고 한다. 15세가 되자 갈루아는 라그랑주의 《대수방정식 해법 탐구》를 읽을 정도로 수학에 재능이 있었다.

1828년 갈루아는 더 높은 수준의 수학 공부를 하기 위해서 프랑스 최고의 수학과가 있었던 에콜 폴리테크니크 입학시험에 응시했다. 그러나 구두시험에 대한 설명이 서툴러 면접시험에서 떨어졌다. 결국 그는 당시 수학 과목의 평판이 형편없었던 에콜 노르말 쉬페리외르에 진학하게 된다.

갈루아의 아버지는 짧은 풍자시를 짓고는 했다. 반대파는 그것을 교묘하게 위조해서 갈루아의 아버지를 공격하고, 시장직에서 물러나게 했다. 1829년 7월 28일, 갈루아의 아버지는 수치심을 이겨내지 못하고 자살했다. 갈루아가 에콜 노르말에 입학한 것은 아버지가 죽은 지 며칠 지나지 않아서였다.

1829년 갈루아는 연분수를 주제로 한 첫 번째 논문을 발표하는 동시에 다항방정식을 연구하고 있었다. 그는 프랑스 과학 아카데미 앞으로 두 편의 논문을 보냈다. 코시가 갈루아의 논문을 심사했지만, 불명확한 부분을 이유로 출판은 거절했다. 논문에 많은 문제가 있음에도 불구하고, 코시는 갈루아의 연구가 중요한 내용을 담고 있는 것을 알았다. 그래서 아카데미 수학상을 수여할 수 있도록 두 논문을 하나로 합쳐달라고 요청했다. 당대에 가장 유명한 수학자 가운데 한 명이었던 코시는 갈루아가 아카데미 수학상을 수상할 거라고 생각했다.

그렇게 해서 한 편으로 제출한 논문은 아카데미 수학상 심사 위원이었던 푸리에게 전달되었다. 그러나 푸리에가 사망하면서 논문이 분실되었다. 결국 그해 수상자는 야코비였는데, 심사 과정에서 갈루아의 논문과 아벨의 논문은 석연찮은 이유로 제외되었다. 갈루아는 자신의 논문이 심사에서 빠진 것은 정치적 음모 때문이라고 여겼다.

갈루아는 5차 이상의 다항방정식의 해의 존재를 가리는 판별식에 대한 논문을 여러 차례 제출했으나, 살아생전에 출판되지는 못했다. 수상에 실패한 그는 논문을 학회지에 기고하고 출간했다. 첫 번째는 군론에 대한 것이었고, 두 번째는 고차방정식의 해에 대한 것이었으며, 세 번째는 유한체에 대해서 최초로 정리한 것이었다.

갈루아가 살던 시기에 프랑스의 정치 상황은 매우 불안했다. 1824년 루이 18세의 뒤를 이어 샤를 10세가 즉위했다. 하지만 1827년 선거에서 그를 지지하던 정파는 다수당의 지위를 잃었고, 1830년에는 반대파였던 자유당이 다수당이 되었다. 사퇴 압력에 직면한 샤를 10세는 쿠데타를 일으키고 악명 높은 7월 포고령을 선포했다. 그러나 이 때문에 일어난 7월 혁명으로 결국 하야했고, 루이 필리프 1세가 새 왕으로 즉위했다.

에콜 폴리테크니크의 학생들도 혁명에 참여해 거리 곳곳에서 역사를 만들었다. 하지만 이 '영광의 3일' 동안 갈루아와 에콜 노르말의 학생들은 교장에 의해 학교에 감금되어 한 발자국도 나갈 수 없었다. 갈루아는 격분해 교장을 신랄하게 비난하는 편지를 학교 신문에 기고했다. 편지에는 갈루아의 실명이 그대로 담겨 있었다. 학교 신문의 편집진은 갈루아의 이름을 감춘 채 보도했지만, 갈루아는 퇴학당하고 말았다.

편지 사건으로 학교를 그만둔 갈루아는 급진 공화주의 세력이었던 국가방위군 포병대에 가입했다. 그는 수학과 정치 집회에만 몰두했다. 대원들과 함께 정치 토론을 거치면서 갈루아는 자연스레 국가방위군 포병대의 일원이 되었다. 1830년 12월 31일, 국가방위군 포병대는 정부를 약화시키고자 한다는 이유로 해산되었다. 같은 시기 포병대 장교였던 19세의 갈루아는 국가 전복 음모에 가담했다는 혐의로 체포되었다.

1831년 4월이 되자 포병대 장교들에게 씌워진 혐의가 벗겨져 모두 석방되었다. 갈루아는 바스티유의 날인 7월 14일, 국가방위군 포병대 제복을 입고 권총과 소총, 단도로 무장한 채 공화주의 시위를 주도했다. 이 일로 그는 체포되었으며, 금지된 국가방위군 포병대 군복을 입었다는 이유로 6개월 금고형을 선고받았다.

갈루아는 에콜 노르말 쉬페리외르에서 퇴학당한 이후 혼자서 대수학 공부를 이어갔다. 하지만 언제나 정치적 행동이 먼저였다. 1832년 1월, 갈루아는 푸아송의 요청으로 방정식 이론에 대한 논문을 제출했다. 푸아송은 7월이 되어서야 "갈루아의 방정식 이론은 충분히 엄밀하지도 않고 더 이상 엄밀하게 발전시킬 수도 없어서 이해할 수 없다"며 거절했다. 갈루아는 이 평가서를 보고 매우 격하게 반발했다. 그는 차라리 자비로 출간할지언정 아카데미를 통해서 논문을 출판하는 것은 단념하겠다고 다짐했다.

갈루아는 석연찮은 이유로 1832년 5월 30일 수요일 아침에 결투를 하게 되었다. 기록에 의하면 명사수 데르벵빌은 스테파니와 갈루아가 사랑에 빠졌음을 눈치채고 갈루아에게 결투를 청했다. 갈루아가 왜 결투를 피하지 않았는지에 대해서는 추측만 무성할 뿐이다. 결투 5일 전 갈루아는 친구 슈발리에에게 이루어지지 않은 사랑 때문에 결투를 하게 되었다고 편지를 썼다. 1832년 5월 30일, 갈루아는 권총을 사용한 결투에서 오른쪽 복부에 총상을 입었다. 총알은 내장을 뚫고 들어가 왼쪽 둔부에 박혔다. 다음 날 갈루아는 사망했고, 6월 1일 사망확인서가 발급되었다.

19세기 권총 결투

군의 정의 _ 이항연산과 네 가지 조건

정교수 이제 갈루아가 창시한 군론 이야기를 해볼게. 군론 아이디어는 그 이전에 가우스, 라그랑주, 루피니, 코시 등이 먼저 생각했지만, 군을 수학에서 하나의 대상으로 처음 다룬 것은 갈루아야. 그래서 갈루아를 군론의 창시자라고 불러.

군론에 들어가기 앞서 이항연산(binary operation)을 알아보자. 이항연산은 두 수를 하나의 수에 대응시키는 것을 말한다. 사칙연산이 대표적인 예이다. 우선 덧셈(+)을 생각해 보자. 2+3=5라는 연산은 두 수 2와 3을 하나의 수 5에 대응시키는 과정이므로 덧셈은 이항연산이다.

사칙연산 중에서 덧셈과 곱셈에 대해서는 교환법칙이 성립한다.

$a + b = b + a$

$ab = ba$

하지만 뺄셈과 나눗셈에 대해서는 교환법칙이 성립하지 않는다. 그리고 덧셈과 곱셈에 대해서는 결합법칙이 성립하지만, 뺄셈과 나눗셈에 대해서는 결합법칙이 성립하지 않는다.

$(a + b) + c = a + (b + c)$

$(ab)c = a(bc)$

일반적인 이항연산을 $*$로 쓰자. 두 수 a, b의 이항연산은 $a*b$라고 쓴다. 이항연산 $*$에 대해 교환법칙과 결합법칙이 성립한다면

$$a * b = b * a$$

$$a * (b * c) = (a * b) * c$$

가 된다. 예를 들어 다음과 같은 이항연산을 만들어 보자.

$$a * b = a + b + ab$$

이 연산은 교환법칙과 결합법칙을 모두 만족한다.

이번에는 '닫혀 있다'의 개념을 알아보자. 어떤 집합 A를 생각하자. 이 집합의 원소 중 임의로 두 개를 선택해서 이항연산 $*$를 적용한 것이 다시 이 집합에 속하는 원소가 될 때, 집합 A는 이항연산 $*$에 대해 닫혀 있다고 말한다.

집합 A = {1, 2, 3}이라고 하자. 이항연산을 덧셈으로 생각할 때, 집합 A는 덧셈에 대해 닫혀 있는지 알아보자. 예를 들어 2와 3을 선택해 덧셈을 하면 5가 되는데, 5는 집합 A의 원소가 아니다. 임의의 두 원소는 같은 두 개를 선택할 수도 있다. 그러니까 가능한 모든 경우는 다음과 같다.

$$1 + 1 = 2$$

$$1 + 2 = 2 + 1 = 3$$

$2 + 2 = 4$

$1 + 3 = 3 + 1 = 4$

$2 + 3 = 3 + 2 = 5$

$3 + 3 = 6$

두 원소를 선택해서 덧셈을 하면 집합 A의 원소가 되기도 하고 안 되기도 한다. 이런 경우 이 집합은 덧셈에 대해 닫혀 있지 않다고 말한다.

이번에는 자연수의 집합을 보자.

$N = \{1, 2, 3, 4, \cdots\}$

이 집합은 덧셈과 곱셈에 대해 닫혀 있다.

자연수의 집합을 다시 보자. 이항연산으로는 곱셈을 생각하자. 1은 자연수인데 1과 어떤 자연수 a의 곱은 항상 a이다. 즉,

$1 \times a = a$

이다. 이렇게 이항연산을 했을 때 자기 자신이 나오게 하는 원소를 항등원이라고 부른다. 즉, 자연수의 집합에서 곱셈에 대한 항등원은 1이다.

이것을 일반적인 이항연산 * 에 대해 생각하면 다음과 같이 말할 수 있다. 집합 A에 속하는 임의의 원소 a에 대해

$$a * e = e * a = a$$

를 만족하는 e를 집합 A에서 이항연산 $*$에 대한 항등원이라고 부른다.

자연수의 집합에서 덧셈에 대한 항등원은 없다. 만약 덧셈에 대한 항등원을 e라고 하면, 임의의 자연수 a에 대해

$$a + e = e + a = a$$

가 되어야 한다. 이것은 $e = 0$을 의미하는데 0은 자연수 집합의 원소가 아니다. 따라서 자연수의 집합은 덧셈에 대한 항등원이 존재하지 않는다. 반면 0은 정수 집합의 원소이므로 정수 집합은 곱셈에 대한 항등원과 덧셈에 대한 항등원을 모두 가지고 있다.

이제 역원을 알아보자. 집합 A의 원소 a에 이항연산 $*$를 적용했을 때, 항등원이 나오게 하는 A의 원소를 a의 역원이라고 부른다. 예를 들어 정수의 집합에서 이항연산으로 덧셈을 생각해 보자. 이때 항등원은 0이므로 정수 집합의 원소 1의 역원은 −1이다.

$$1 + (-1) = 0$$

일반적으로 정수의 집합에서 원소 a의 역원은 $-a$이다. 자연수의 집합은 덧셈에 대한 항등원이 없으니까 역원을 정의할 수 없다. 마찬가지로 정수의 집합에서 이항연산을 곱셈으로 생각하면 역원이 안 생기는 경우도 있다. 예를 들어 곱셈의 항등원은 1인데 2의 역원을 x라고 하면

$$2x = 1$$

이어야 한다. 이것을 풀면 $x = \frac{1}{2}$이 된다. 이러한 원소는 정수의 집합에 속하지 않으므로 정수의 집합에서 2의 곱셈에 대한 역원은 존재하지 않는다.

지금까지 공부한 내용을 토대로 우리는 군이라는 개념을 만들 수 있다. 군을 정의하려면 집합과 어떤 연산이 필요하다. 집합 A와 이항연산 *에 대해 다음과 같은 네 개의 조건을 보자.

[조건 1] 집합 A는 이항연산 *에 대해 닫혀 있다.
[조건 2] 이항연산 *는 결합법칙을 만족한다.
[조건 3] 집합 A는 이항연산 *에 대한 항등원 e를 갖는다.
[조건 4] 집합 A의 모든 원소는 이항연산 *에 대한 역원을 갖는다.

이 네 조건을 만족하는 집합 A는 이항연산 *에 대해 군을 이룬다고 말한다.

군을 정의할 때 이항연산이 반드시 교환법칙을 만족할 필요는 없다. 특별히 이항연산이 교환법칙을 만족할 때 이 군을 아벨군이라고 부른다.

이제 어떤 집합이 군이 되는지 알아보자. 가장 간단한 군은 다음과 같다.

$$A = \{1\}$$

이 집합은 곱셈에 대해 군을 이룬다. 이 집합은 원소가 하나이므로 두 개를 뽑아 곱하면 1 × 1 = 1이다. 즉, 이 집합은 곱셈에 대해 닫혀 있으니까 조건 1을 만족한다. 연산을 곱셈으로 택했고 곱셈은 결합법칙이 성립하므로 조건 2도 만족한다. 이 집합은 곱셈에 대한 항등원 1을 원소로 가지므로 조건 3을 만족한다. 1의 곱셈에 대한 역원은 1이니까 조건 4도 만족한다. 따라서 이 집합은 조건 1~4를 모두 만족하므로 곱셈을 연산으로 하는 군이다. 곱셈은 교환법칙이 성립하니까 이 군은 아벨군이다. 하지만 이 집합은 덧셈에 대해서는 군을 이루지 않는다.

이번에는 다음 집합 B를 보자.

B = {−1, 1}

이 집합은 덧셈에 대해 군을 이루지 않지만, 곱셈에 대해서는 군을 이룬다.

회전군 _ 반시계 방향으로 몇 도 회전시킬까?

정교수   이번에는 회전군에 대해 알아볼게.

다음 그림을 보자.

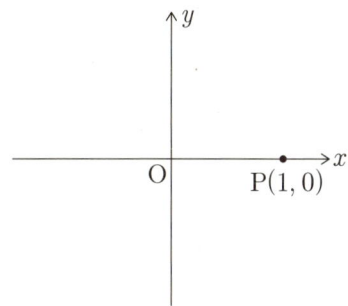

이제 다음과 같은 네 개의 작용을 생각하자.

$R(0)$: 반시계 방향으로 0도 회전
$R(90)$: 반시계 방향으로 90도 회전
$R(180)$: 반시계 방향으로 180도 회전
$R(270)$: 반시계 방향으로 270도 회전

점 P에 $R(0)$을 작용하면 점 P가 된다. 이번에는 점 P에 $R(90)$을 작용하자. 그러면 점 P는 다음과 같이 점 P′으로 이동한다.

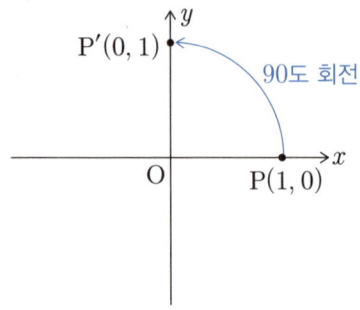

마찬가지로 점 P에 $R(180)$을 작용하면 점 P는 다음과 같이 점 P″으로 이동한다.

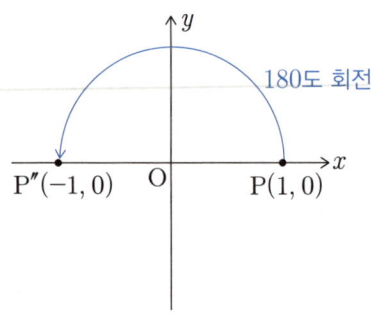

같은 방법으로 점 P에 $R(270)$을 작용하면 점 P는 다음과 같이 점 P‴으로 이동한다.

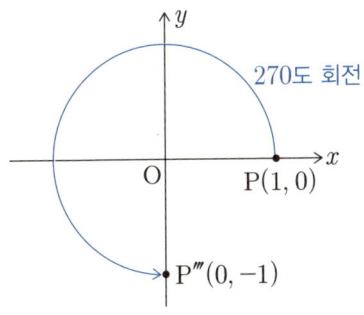

이렇게 네 개의 작용을 원소로 갖는 집합을 G라고 하자.

G = {R(0), R(90), R(180), R(270)}

이제 이 작용들 사이의 연산을 정의하자. 예를 들어 두 원소 R(90) 과 R(180)의 연산을 다음과 같이 쓰자.

R(90) * R(180)

이 연산은 * 뒤에 있는 것을 먼저 작용하고 * 앞에 있는 것을 나중에 작용하는 것으로 약속하자. 다음 그림을 보자.

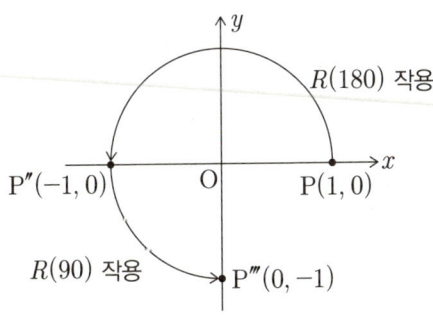

이때 점 P가 점 P‴으로 이동한다. 즉, 이 연산의 결과는 270도 회전한 것과 같다.

$R(90) * R(180) = R(270)$

먼저 90도 회전하고 나중에 180도 회전해도 같은 결과가 나온다.

$R(180) * R(90) = R(270)$

이번에는 먼저 180도 회전하고 나중에 270도 회전하는 경우를 보자. 이것을 그림으로 그리면 다음과 같다.

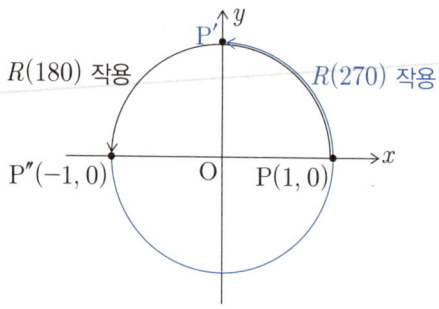

점 P가 점 P′으로 이동하므로 이 연산의 결과는 90도 회전한 것과 같다.

이제 집합 G가 연산 *에 대해 군을 이루는지 조사하자. 먼저 조건 1을 확인하자. 가능한 모든 연산 결과를 나열하면 다음과 같다.

$R(0) * R(0) = R(0)$

$R(0) * R(90) = R(90)$

$R(0) * R(180) = R(180)$

$R(0) * R(270) = R(270)$

$R(90) * R(0) = R(90)$

$R(90) * R(90) = R(180)$

$R(90) * R(180) = R(270)$

$R(90) * R(270) = R(0)$

$R(180) * R(0) = R(180)$

$R(180) * R(90) = R(270)$

$R(180) * R(180) = R(0)$

$R(180) * R(270) = R(90)$

$R(270) * R(0) = R(270)$

$R(270) * R(90) = R(0)$

$R(270) * R(180) = R(90)$

$R(270) * R(270) = R(180)$

모든 연산 결과가 집합 G의 원소가 되므로 집합 G는 연산 *에 대해 닫혀 있다. 앞의 결과를 자세히 들여다보면 *는 교환법칙을 만족하는 것을 알 수 있다.

이제 조건 2를 확인하자. 다음 경우를 보자.

$(R(90) * R(180)) * R(270)$

괄호 안을 먼저 셈해야 하므로

$(R(90) * R(180)) * R(270) = R(270) * R(270) = R(180)$

이 된다. 이번에는 다음 식을 보자.

$R(90) * (R(180) * R(270)) = R(90) * R(90) = R(180)$

따라서 우리는 다음 결과를 얻는다.

$(R(90) * R(180)) * R(270) = R(90) * (R(180) * R(270))$

즉, 연산 *에 대해 결합법칙이 성립한다.
이제 연산 *에 대한 항등원을 찾아보자.

$R(0) * R(0) = R(0)$

$R(90) * R(0) = R(90)$

$R(180) * R(0) = R(180)$

$R(270) * R(0) = R(270)$

따라서 항등원은 $R(0)$이고, 조건 3을 만족한다.

이제 집합 G의 모든 원소에 대한 역원을 찾아보자. 조건 1에서 나열한 연산 결과가 항등원이 되는 식을 골라보면 다음과 같다.

$R(0) * R(0) = R(0)$

$R(90) * R(270) = R(0)$

$R(180) * R(180) = R(0)$

$R(270) * R(90) = R(0)$

이로부터 다음 사실을 알 수 있다.

$R(0)$의 역원 $= R(0)$
$R(90)$의 역원 $= R(270)$
$R(180)$의 역원 $= R(180)$
$R(270)$의 역원 $= R(90)$

집합 G의 모든 원소에 대한 역원이 존재하므로 조건 4를 만족한다. 집합 G는 연산 * 와 함께 조건 1~4를 만족하기 때문에 군을 이룬다. 물론 아벨군이며, 이 군을 수학자들은 회전군이라고 부른다.

마리우스 솝후스 리 _동료들과 함께 학문적 열정을 쏟다

정교수 이제 리군의 창시자인 마리우스 솝후스 리에 대해 알아볼게.

마리우스 솝후스 리(Marius Sophus Lie, 1842~1899)

마리우스 솝후스 리는 노르웨이 노르피오르데이드의 작은 마을에서 태어났다. 그는 루터교 목사인 아버지 요한 헤르만 리(Johann Herman Lie)와 노르웨이 트론헤임의 명문 출신 어머니 사이에서 육 남매 중 막내로 태어났다.

솝후스 리는 남부 노르웨이의 모스에서 초등교육을 받은 후, 당시 '크리스티아니아'로 불렸던 오슬로에서 고등학교를 다녔다. 고등학교 졸업 후에는 군인이 되기를 꿈꿨으나, 시력 문제로 입대를 거부당해 꿈은 좌절되었다. 그는 크리스티아니아 대학(현재 오슬로 대학)에 진학해 수학 공부를 이어갔다.

크리스티아니아 대학(출처: Ryan Hodnett/Wikimedia Commons)

　솝후스 리의 첫 번째 수학 논문인 〈평면기하학에서 허수의 표현〉은 1869년 크리스티아니아 과학 아카데미와 독일의 《크렐레지(Crelle's Journal)》에 발표되었다. 같은 해 그는 장학금을 받아 베를린으로 유학했고, 1869년 9월부터 1870년 2월까지 그곳에 머물렀다. 이때 그는 훗날 유명한 수학자가 되는 펠릭스 클라인(Felix Klein)을 만나 절친한 사이가 되었다.

　베를린을 떠난 솝후스 리는 파리로 향했고, 두 달 뒤 클라인도 그와 합류했다. 그들은 그곳에서 카미유 조르당(Camille Jordan), 가스통 다르부(Gaston Darboux) 등 당대 수학자들과 교류했다. 그러나 1870년 7월 19일, 프로이센-프랑스 전쟁이 발발하면서 상황이 급변했다. 프로이센 국적이었던 클라인은 프랑스를 급히 떠나야 했고, 솝후스 리는 퐁텐블로에서 독일 스파이로 오해받아 체포되기에 이른다. 솝후스 리는 한때 감옥에 수감되었으나, 다르부의 중재 덕분에 한

달 뒤 석방되었다. 이 사건은 노르웨이에서도 큰 관심을 끌며 솝후스 리의 명성을 높이는 계기가 되었다.

1871년 솝후스 리는 크리스티아니아 대학에서 〈기하학적 변환의 한 종류에 대하여〉라는 논문으로 박사 학위를 받았다. 이 논문은 훗날 프랑스 수학자 가스통 다르부에 의해 "현대 기하학의 가장 멋진 발견 중 하나"로 평가받으며 큰 반향을 일으켰다.

솝후스 리의 업적은 곧 노르웨이 사회 전반에서 인정받았다. 1872년에는 노르웨이 의회가 그를 위한 특별한 수학 교수직을 신설하는 전례 없는 조치를 취했다. 같은 해 그는 독일 에를랑겐을 방문해 당시 '에를랑겐 프로그램'을 주도하고 있던 친구 펠릭스 클라인을 다시 만났다. 이해는 또한 솝후스 리에게 있어 수학사적으로 중요한 시간이었다. 그는 동료 수학자 페테르 루드비 실로우(Peter Ludwig Sylow)와 함께 그들의 선배이자 노르웨이의 수학 영웅 닐스 헨리크 아벨(Niels Henrik Abel)의 연구 내용을 편집하고 출판하는 데 헌신했다.

1872년 말, 솝후스 리는 18세의 안나 버치(Anna Birch)에게 청혼했고, 이들은 1874년에 결혼했다. 부부는 세 자녀를 두었는데 각각 마리(1877년생), 다그니(1880년생), 헤르만(1884년생)이다.

수학 외에도 솝후스 리는 학문 공동체를 형성하는 데 적극적으로 참여했다. 1876년부터 그는 의사 야콥 보름-뮐러(Jacob Worm-Müller), 생물학자 게오르그 오시안 사르스(Georg Ossian Sars)와 함께 노르웨이 과학 학술지 《Mathematik og Naturvidenskab》의 공동 편집을 맡았다.

1884년 독일 수학자 프리드리히 엥겔(Friedrich Engel)이 클라인과 아돌프 마이어의 추천으로 크리스티아니아를 찾아와 쇼후스 리의 연구를 돕게 되었다. 이 협력은 곧 쇼후스 리의 대표작인 《변환군 이론》으로 결실을 맺는다. 이 책은 1888년부터 1893년까지 세 권으로 라이프치히에서 출판되었다. 엥겔은 이 방대한 작업에서 공동 저자이자 실질적인 조력자로 활약했다. 그는 훗날 쇼후스 리의 유고 논문집 편집에도 참여했다.

1886년 클라인이 괴팅겐으로 자리를 옮기자 쇼후스 리는 그의 뒤를 이어 라이프치히 대학의 교수로 임명되었다. 그러나 학문적 열정과 달리 건강은 점점 악화했다. 1889년 11월, 쇼후스 리는 심각한 정신쇠약 증세를 보이며 1890년 6월까지 병원에 입원해야 했다. 이후 다시 연구와 강단에 복귀했지만 빈혈 증세가 점차 심해졌다. 그는 결국 1898년 5월에 교수직을 사임하고, 그해 9월 고국 노르웨이로 귀환했다. 이듬해인 1899년, 그는 비타민 B12 흡수 장애로 인한 악성 빈혈로 56세의 나이에 생을 마감했다.

행렬군과 리군 _ 행렬군이 특별한 관계를 만족할 때

정교수 먼저 행렬군을 알아볼게.
물리군 모든 행렬이 군을 이루나요?
정교수 그렇지는 않아. 역행렬이 존재하는 정사각행렬의 집합이 군을 이루는데, 이 군을 행렬군이라고 불러.

일반적인 행렬군은 다음과 같이 정의한다.

$$G = \left\{ \begin{pmatrix} a & b \\ c & d \end{pmatrix} \mid ad - bc \neq 0 \right\}$$

집합 G는 행렬 곱셈에 대해 닫혀 있으니까 군의 조건 1을 만족하고, 행렬 곱셈은 결합법칙을 만족하므로 군의 조건 2를 만족한다. 이 집합은 항등원

$$\begin{pmatrix} 1 & 0 \\ 0 & 1 \end{pmatrix}$$

을 가지므로 군의 조건 3을 만족한다. 또한 $\begin{pmatrix} a & b \\ c & d \end{pmatrix}$의 역원은

$$\frac{1}{ad-bc} \begin{pmatrix} d & -b \\ -c & a \end{pmatrix}$$

가 되고, 이 행렬이 존재하므로 군의 조건 4를 만족한다.

물리군 행렬군과 리군이 관계있나요?

정교수 물론이야. 솝후스 리는 행렬군 중에서 행렬이 어떤 특별한 관계를 만족하는 군들을 연구했어. 이것을 솝후스 리의 이름을 따서 리군이라고 불러.

물리군 리군에는 여러 종류가 있겠네요.

정교수 맞아. 간단한 리군을 소개할게. 2차원 평면에서 위치벡터를 그려볼까?

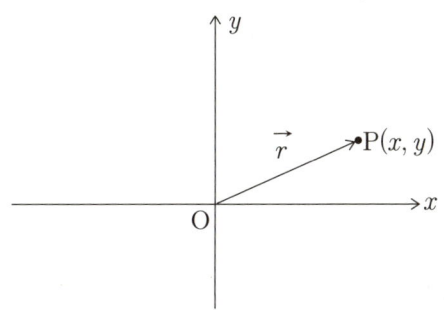

위 그림에서 점 P(x, y)의 위치벡터 $\vec{r}$을 성분표시로 나타내면

$$\vec{r} = (x, y)$$

이다. 이것을 다음과 같이 행렬 X에 대응시켜 보자.

$$X = \begin{pmatrix} x \\ y \end{pmatrix}$$

그러면 행렬 X로 점 P를 묘사할 수 있다. 이때 원점에서 점 P까지의 거리를 r이라고 하면

$$r^2 = x^2 + y^2$$

이다. 행렬의 행과 열을 바꾸는 것을 전치라고 하는데, 행렬 X의 전치

를 X^T로 쓰면

$$X^T = (x \ y)$$

가 된다. 그러니까

$$r^2 = X^T X$$

이다.

솝후스 리는 점 P가 원점을 중심으로 θ만큼 회전해 점 P′이 되는 회전변환을 생각했다. 점 P의 좌표를 (x, y), 이동한 점 P′의 좌표를 (x', y')으로 두고 그림으로 나타내면 다음과 같다.

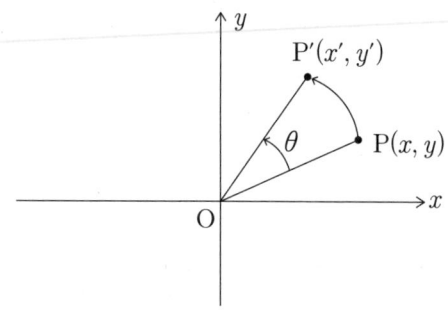

원점을 중심으로 회전시켰으므로 원점으로부터 두 점까지의 거리는 달라지지 않는다. 즉,

$$\overline{OP'} = \overline{OP}$$

이고, 이것은

$$x'^2 + y'^2 = x^2 + y^2$$

을 의미한다.

$$X' = \begin{pmatrix} x' \\ y' \end{pmatrix}$$

으로 나타내면

$$X'^T X' = X^T X \qquad\qquad (2\text{-}5\text{-}1)$$

이다.

다음 그림과 같이 선분 OP와 x축의 양의 방향이 이루는 각을 α라고 하자.

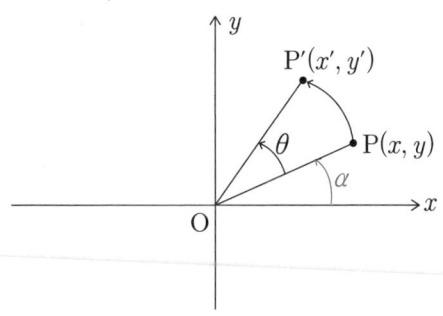

이때

$$x = r\cos\alpha$$

$$y = r\sin\alpha$$

이고, 마찬가지로

$$x' = r\cos(\theta + \alpha)$$

$$y' = r\sin(\theta + \alpha)$$

이다. 삼각함수의 덧셈정리를 이용하면

$$x' = r[\cos\theta\cos\alpha - \sin\theta\sin\alpha]$$

이므로

$$x' = x\cos\theta - y\sin\theta$$

가 되고, 마찬가지로

$$y' = x\sin\theta + y\cos\theta$$

이다.

이것을 행렬로 나타내면

$$X' = R(\theta)X$$

이고, 여기서

$$R(\theta) = \begin{pmatrix} \cos\theta & -\sin\theta \\ \sin\theta & \cos\theta \end{pmatrix}$$

이다. 이 행렬은 θ만큼 회전시키는 변환을 하는 행렬이므로 회전행렬이라고 부른다.

물리군 회전행렬들이 군을 이루나요?

정교수 물론이야. 이 군을 G라고 해볼까? θ만큼 회전시키는 변환을 한 후 θ'만큼 회전시키는 변환을 하는 것은 $\theta + \theta'$만큼 회전시키는 변환을 하는 것과 마찬가지야. 그러니까

$$R(\theta')R(\theta) = R(\theta + \theta') \in G$$

가 되어 조건 1을 만족하고, 행렬 곱셈은 결합법칙이 성립하므로 조건 2를 만족해. 그리고 항등원은

$$R(0) = \begin{pmatrix} 1 & 0 \\ 0 & 1 \end{pmatrix}$$

이고, $R(\theta)$의 역원은 $R(-\theta)$이지. 따라서 2차원에서 회전변환을 시키는 변환행렬들은 군을 이뤄.

솝후스 리는 어떤 변환을 통해 원점으로부터의 거리를 변하지 않게 하는 변환행렬들의 모임이 군을 이룰 때, 이를 직교군이라고 불렀어. 2차원에서의 직교군은 $O(2)$로 나타내. 그러니까

$$X' = AX \tag{2-5-2}$$

라는 변환이

$$X'^T X' = X^T X \tag{2-5-3}$$

를 만족하게 하는 모든 A가 $O(2)$군을 만들지. 전치의 성질에 의해

$$X'^T = (AX)^T = X^T A^T$$

이므로 식 (2-5-3)으로부터

$$X^T A^T A X = X^T X$$

가 되어

$$A^T A = I \tag{2-5-4}$$

임을 알 수 있어. 즉, $O(2)$군은 식 (2-5-4)를 만족하는 행렬들의 집합이야. 솝후스 리는 $O(2)$군의 모든 행렬에 대해 행렬식의 값이 1일 때, 이들 행렬로 이루어진 $O(2)$군은 특별하니까 special의 S를 따서 $SO(2)$군이라고 불렀어.

물리군 2차원 회전행렬들은 $SO(2)$군을 이루나요?

정교수 $R(\theta)$의 행렬식의 값은 1이니까 2차원 회전행렬들은 $SO(2)$군을 이루지.

물리군 그렇다면 3차원에서 회전행렬들은 $SO(3)$군을 이루겠네요.

정교수 맞아. 3차원에서 회전행렬들은 $SO(3)$군의 원소야.

x축 주위로의 회전각을 θ_1이라고 하면 x축 주위로의 회전을 나타내는 $SO(3)$군의 원소는

$$R_x(\theta_1) = \begin{pmatrix} 1 & 0 & 0 \\ 0 & \cos\theta_1 & -\sin\theta_1 \\ 0 & \sin\theta_1 & \cos\theta_1 \end{pmatrix}$$

이다. y축 주위로의 회전각을 θ_2라고 하면 y축 주위로의 회전을 나타내는 $SO(3)$군의 원소는

$$R_y(\theta_2) = \begin{pmatrix} \cos\theta_2 & 0 & \sin\theta_2 \\ 0 & 1 & 0 \\ -\sin\theta_2 & 0 & \cos\theta_2 \end{pmatrix}$$

이다. z축 주위로의 회전각을 θ_3이라고 하면 z축 주위로의 회전을 나타내는 $SO(3)$군의 원소는

$$R_z(\theta_3) = \begin{pmatrix} \cos\theta_3 & -\sin\theta_3 & 0 \\ \sin\theta_3 & \cos\theta_3 & 0 \\ 0 & 0 & 1 \end{pmatrix}$$

이다. $SO(3)$군의 일반적인 원소는

$$R_x(\theta_1)R_y(\theta_2)R_z(\theta_3)$$

이 된다.

유니터리군 _ N차원 복소 공간에서

정교수   4차원 공간을 생각해 볼까? 4차원 공간의 한 점은

$$(x_1, x_2, x_3, x_4)$$

로 묘사할 수 있어. 여기서 x_1, x_2, x_3, x_4는 실수야. 이때 4차원 공간에서 원점 (0, 0, 0, 0)과 이 점 사이의 거리를 L이라고 하면

$$L^2 = x_1^2 + x_2^2 + x_3^2 + x_4^2$$

이 돼. 이제 다음과 같이 복소수를 도입할게.

$$z_1 = x_1 + ix_2$$

$$z_2 = x_3 + ix_4$$

물리군   그럼 4차원에서 원점으로부터의 거리의 제곱은

$$z_1^2 + z_2^2$$

인가요?

정교수   그렇지 않아.

$$z_1^2 + z_2^2 = x_1^2 - x_2^2 + 2x_1x_2 i + x_3^2 - x_4^2 + 2x_3x_4 i$$

는 복소수야. 복소수가 두 점 사이의 거리의 제곱이 될 수는 없어.

물리군 그렇군요.

정교수 이때는 켤레복소수를 도입하면 돼. 두 복소수 z_1, z_2의 켤레는

$$z_1^* = x_1 - ix_2$$

$$z_2^* = x_3 - ix_4$$

이니까

$$L^2 = z_1^* z_1 + z_2^* z_2$$

이지.

이제 2차원 복소 공간의 한 점을 나타내는 (z_1, z_2)를 다음과 같이 열벡터 Z로 나타내자.

$$Z = \begin{pmatrix} z_1 \\ z_2 \end{pmatrix}$$

이 행렬의 각 성분에 켤레를 취해 전치시킨 행렬을 Z의 수반이라 하고 다음과 같이 나타내자.

$$Z^\dagger = (z_1^* \ z_2^*)$$

즉,

$$Z^\dagger = (Z^T)^*$$

이다. 그러므로 2차원 복소 공간에서 원점으로부터 (z_1, z_2)까지의 거리의 제곱은

$$L^2 = Z^\dagger Z$$

이다. 전치의 성질로부터 수반에 대한 다음과 같은 성질을 알 수 있다.

(1) $(A^\dagger)^\dagger = A$

(2) $(AB)^\dagger = B^\dagger A^\dagger$

2차원 복소 공간에서 거리의 제곱 L^2이 바뀌지 않게 하는 변환을 찾아보자. 먼저 두 열벡터 Z, Z' 사이에 다음과 같은 변환이 있다고 하자.

$$Z' = AZ$$

여기서 A는 변환행렬이다. 변환 후에도 원점으로부터의 거리가 같으려면

$$Z'^\dagger Z' = Z^\dagger Z$$

를 만족해야 한다. 즉,

$$(AZ)^\dagger (AZ) = Z^\dagger Z$$

이고, 수반의 성질을 이용하면

$$Z^\dagger A^\dagger A Z = Z^\dagger Z$$

이므로

$$A^\dagger A = I \qquad (2\text{-}6\text{-}1)$$

또는

$$A^\dagger = A^{-1} \qquad (2\text{-}6\text{-}2)$$

이다. 식 (2-6-1)을 만족하는 행렬을 유니터리(unitary)행렬이라고 부르자. 2차원 복소 공간에서 유니터리행렬로 이루어진 군을 $U(2)$군이라 하고, 이 중 모든 원소의 행렬식의 값이 1인 원소들로만 이루어진 군을 $SU(2)$군이라고 부르자. 여기서 SU는 'special unitary'의 머리글자다.

이제 $SU(2)$군의 원소를 구해보자. 임의의 원소를

$$A = \begin{pmatrix} a & b \\ c & d \end{pmatrix}$$

라고 하면 여기서 a, b, c, d는 복소수이다. 행렬식이 1이므로

$$ad - bc = 1$$

이다. 식 (2-6-2)로부터

$$\begin{pmatrix} a^* & c^* \\ b^* & d^* \end{pmatrix} = \begin{pmatrix} d & -b \\ -c & a \end{pmatrix}$$

이므로

$$c = -b^*$$

$$d = a^*$$

가 된다. 따라서 $SU(2)$군의 원소는

$$A = \begin{pmatrix} a & b \\ -b^* & a^* \end{pmatrix}$$

이고

$$|a|^2 + |b|^2 = 1$$

이 된다. 여기서 a, b는 복소수이므로 a는 2개의 실수로 묘사되고, 마찬가지로 b도 2개의 실수로 묘사된다. 그런데 이 4개의 실수 사이에 $|a|^2 + |b|^2 = 1$이 성립하므로 독립적인 실수의 수는 3개이다. 즉, $SU(2)$군의 원소의 개수는 3개이다.

일반적으로 N차원 복소 공간에서 원점과 어떤 점 사이의 거리를 불변하게 하는 변환행렬이 이루는 군은 $U(N)$군이다. 이 중 행렬식의 값이 1인 원소만으로 이루어진 군은 $SU(N)$군이다. $SU(N)$군의 원소의 개수는 $N^2 - 1$(개)이다. 따라서 $SU(3)$군의 원소의 개수는 8개이다.

리대수 _ 변수가 아주 작은 경우

정교수 솝후스 리는 리군으로부터 리대수라는 개념을 처음 도입했어.
물리군 그건 뭐예요?
정교수 리군의 원소에서 변수가 아주 작은 경우를 고려하는 거야. 회전행렬에서 회전 각도가 아주 작은 경우를 생각하는 거지.

$SO(3)$군에서 $\theta_1, \theta_2, \theta_3$이 아주 작은 경우를 생각하자. 테일러 전개에 의해

$$\cos\theta_i \approx 1$$

$$\sin\theta_i \approx \theta_i$$

이므로 이것을 이용하면 다음과 같이 근사시킬 수 있다.

$$R_x(\theta_1) \approx \begin{pmatrix} 1 & 0 & 0 \\ 0 & 1 & -\theta_1 \\ 0 & \theta_1 & 1 \end{pmatrix} = I + \theta_1 L_1$$

$$R_y(\theta_2) \approx \begin{pmatrix} 1 & 0 & \theta_2 \\ 0 & 1 & 0 \\ -\theta_2 & 0 & 1 \end{pmatrix} = I + \theta_2 L_2$$

$$R_z(\theta_3) \approx \begin{pmatrix} 1 & -\theta_3 & 0 \\ \theta_3 & 1 & 0 \\ 0 & 0 & 1 \end{pmatrix} = I + \theta_3 L_3$$

여기서 I는 단위행렬이고

$$L_1 = \begin{pmatrix} 0 & 0 & 0 \\ 0 & 0 & -1 \\ 0 & 1 & 0 \end{pmatrix}$$

$$L_2 = \begin{pmatrix} 0 & 0 & 1 \\ 0 & 0 & 0 \\ -1 & 0 & 0 \end{pmatrix}$$

$$L_3 = \begin{pmatrix} 0 & -1 & 0 \\ 1 & 0 & 0 \\ 0 & 0 & 0 \end{pmatrix}$$

이다. L_1, L_2, L_3은 미소 회전변환을 생성하는데, 솝후스 리는 이 세 행렬을 생성자라고 불렀다. 그는 또한 두 행렬에 대한 교환자를 다음과 같이 정의했다.

$[X, Y] = XY - YX$

이때 세 개의 생성자 L_1, L_2, L_3은

$[L_1, L_2] = L_3$

$[L_2, L_3] = L_1$

$[L_3, L_1] = L_2$

를 만족하는데 이것을 $so(3)$대수라고 부른다.

이제 $SU(2)$군에 대응하는 리대수인 $su(2)$대수를 찾아보자. $SU(2)$ 군의 임의의 원소는 다음과 같이 주어진다.

$$A = \begin{pmatrix} a & b \\ -b^* & a^* \end{pmatrix}$$

여기서

$a = \alpha_0 + i\alpha_3$

$b = \alpha_2 + i\alpha_1$

이라고 두자. 이때 α_0, α_1, α_2, α_3은 실수이다. 이 행렬의 행렬식의 값이 1이므로

$$\alpha_0^2 + \alpha_1^2 + \alpha_2^2 + \alpha_3^2 = 1$$

이 되어

$$\alpha_0 = \sqrt{1 - (\alpha_1^2 + \alpha_2^2 + \alpha_3^2)}$$

이다. 이때

$$A = \sqrt{1 - (\alpha_1^2 + \alpha_2^2 + \alpha_3^2)}\, I + \alpha_1 X_1 + \alpha_2 X_2 + \alpha_3 X_3$$

이고, 세 개의 생성자는

$$X_1 = i\sigma_x$$

$$X_2 = i\sigma_y$$

$$X_3 = i\sigma_z$$

가 된다. 여기서 파울리 행렬을 다음과 같이 정의한다.

$$\sigma_x = \begin{pmatrix} 0 & 1 \\ 1 & 0 \end{pmatrix}$$

$$\sigma_y = \begin{pmatrix} 0 & -i \\ i & 0 \end{pmatrix}$$

$$\sigma_z = \begin{pmatrix} 1 & 0 \\ 0 & -1 \end{pmatrix}$$

파울리 행렬은 다음과 같은 재미있는 성질을 만족한다.

$$\sigma_x^2 = \sigma_y^2 = \sigma_z^2 = I$$

$$\sigma_x \sigma_y = i\sigma_z$$

$$\sigma_y \sigma_x = -i\sigma_z$$

$$\sigma_y \sigma_z = i\sigma_x$$

$$\sigma_z \sigma_y = -i\sigma_x$$

$$\sigma_z \sigma_x = i\sigma_y$$

$$\sigma_x\sigma_z = -i\sigma_y$$

따라서 $su(2)$대수는 세 개의 생성자에 대해 다음과 같은 교환 관계식으로 주어진다.

$[X_1, X_2] = -2X_3$

$[X_2, X_3] = -2X_1$

$[X_3, X_1] = -2X_2$

세 번째 만남

•

새로운 입자의 발견

새로운 입자_입자 발견의 역사

정교수 먼저 물질을 구성하는 입자들이 발견된 역사를 살펴볼게. 최초로 발견된 입자는 전자야. 전자는 1897년 영국의 물리학자 톰슨이 발견했어. 앞으로 전자는 e^-라고 쓸 거야. 전자는 음의 전기를 띠기 때문에 마이너스를 붙였지. 전자의 전하량은 -1이라고 쓰겠네.

톰슨(Sir Joseph John Thomson, 1856~1940, 1906년 노벨 물리학상 수상)

어떤 입자의 질량이 m이라고 하면 이 입자가 가지는 정지에너지는 상대성이론에 의해 mc^2이다. 여기서 c는 광속을 뜻한다.

전자의 정지에너지는 다음과 같다.

0.511MeV

앞으로는 입자의 질량 대신에 정지에너지를 사용하겠다. 정지에너지가 작으면 가벼운 입자를, 정지에너지가 크면 무거운 입자를 의미한다. 입자는 전하량과 질량에 의해 묘사되므로 전자에 대해 다음과 같이 쓰자.

(전자) = (−1, 0.511)

첫 번째 수는 전하량을, 두 번째 수는 정지에너지를 나타낸다.

그다음으로 발견된 입자는 수소의 원자핵인 양성자이다. 양성자는 p로 나타낸다.

(양성자) = (+1, 938.3)

핵을 이루는 또 다른 입자인 중성자는 1932년 채드윅이 발견했다. 중성자는 n으로 나타낸다.

(중성자) = (0, 939.6)

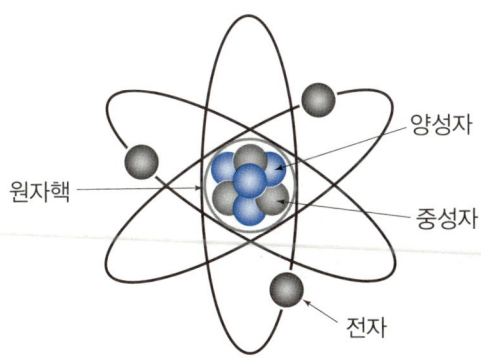

1932년 미국의 앤더슨은 안개상자로 우주 방사선을 연구하던 중 전자와 질량, 전하량의 크기는 같지만 전하의 부호가 반대인 입자를 발견했다. 이 입자는 1928년 영국의 이론물리학자 디랙이 예언했는데, 양의 전기를 띤 전자라는 뜻으로 양전자(positron)라는 이름이 붙었다. 앞으로 양전자를 e^+라고 쓰겠다.

(양전자) = (+1, 0.511)

앤더슨(Carl David Anderson, 1905~1991, 1936년 노벨 물리학상 수상)

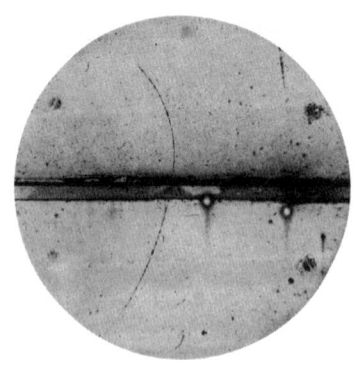

안개상자에서 최초로 관측된 양전자의 궤적

이때부터 모든 입자는 자신의 반입자를 가지는 것이 알려졌다. 이로써 입자의 종류는 두 배로 늘어나게 되었다.

물리군 양성자의 반입자도 발견되었나요?

정교수 물론이야. 1953년 양성자의 반입자인 반양성자를 찾기 위해 두 개의 양성자 가속기가 건설되었어. 하나는 뉴욕 근처 브룩헤이븐 국립연구소의 23억 전자볼트 에너지를 갖는 코스모트론이야. 다른 하나는 캘리포니아 대학의 62억 전자볼트의 베바트론이지. 반양성자는 베바트론에서 발견돼. 1955년 10월, 미국 캘리포니아 대학의 세그레와 체임벌린은 베바트론으로 반양성자를 발견했어. 양성자의 반입자인 반양성자는 $\bar{p}$라고 써.

(반양성자) = (−1, 938.3)

물리군 중성자도 반입자를 가지나요?

정교수 그렇지. 중성자의 반입자를 반중성자라 하고 $\bar{n}$로 쓴다네.

(반중성자) = (0, 939.6)

1956년 로런스 버클리 국립연구소의 브루스 코크(Bruce Cork, 1916~1994) 팀이 양성자-반양성자 충돌 실험에서 반중성자를 발견하지.

브루스 코크(왼쪽)

1936년 미국 캘리포니아 공과대학(Caltech)의 앤더슨(Carl David Anderson)과 네더마이어(Seth Henry Neddermeyer, 1907~1988)는 우주선(cosmic rays)을 연구하던 중 새로운 입자인 뮤온을 발견했다. 뮤온은 음의 전기를 띠므로 μ^-로 나타낸다.

(뮤온) = (−1, 105.7)

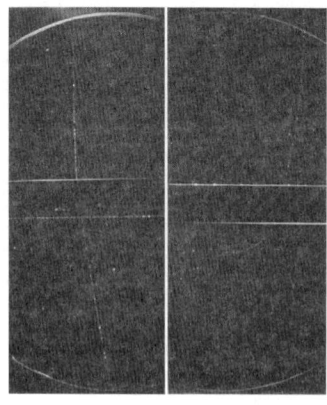

뮤온 입자(출처: Jino john1996/Wikimedia Commons)

뮤온의 반입자를 반뮤온이라 하고 μ^+로 쓴다.

(반뮤온) = (+1, 105.7)

1930년 파울리는 베타 방사선을 내는 방사성물질을 이론적으로 연구하던 중, 전기를 띠지 않고 0에 가까울 정도로 아주 작은 질량을 가진 새로운 입자를 예언했다. 이 입자의 이름은 전자 뉴트리노이다. 전자 뉴트리노는 ν_e로 쓰고, 그것의 반입자인 전자 반뉴트리노는 $\overline{\nu_e}$라고 쓴다.

1953년에 라이너스(Frederick Reines)와 카원(Clyde Cowan)은 뉴트리노 검출에 방해되는 우주선을 피하기 위해 지하 12미터 아래에 실험 장치를 만들었다. 두 사람은 1956년 6월 14일, 전자 뉴트리노를 발견했고 이 소식을 파울리에게 전보로 보냈다. 전자 뉴트리노와 전자 반뉴트리노의 정지에너지는 0.12eV보다 작을 것으로 추정된다.

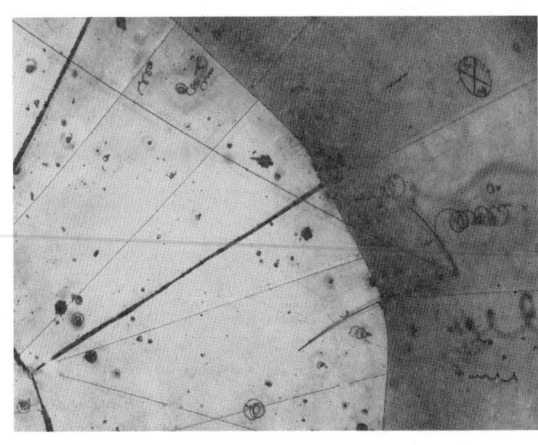

수소 거품상자에서(보이지 않는) 뉴트리노가 양성자와 충돌해(양성자는 중앙 궤도 위의 짧은 선을 따라 이동), 뮤온(긴 중심 직선 트레이스)과 파이온(뮤온 바로 아래의 트레이스)을 생성했다

물리군 전자 뉴트리노는 전자와 관계있는 뉴트리노인가요?

정교수 맞아. 전자 뉴트리노는 베타 방사선이 나올 때 함께 튀어나오는 뉴트리노야. 베타 방사선은 전자들의 흐름이거든. 중성자가 양성자로 바뀌는 베타 붕괴 식은

$$n \rightarrow p + e^- + \overline{\nu_e}$$

가 되어 전자 반뉴트리노가 나타나고, 양성자가 중성자로 바뀌는 베타 붕괴 식은

$$p \rightarrow n + e^+ + \nu_e$$

가 되어 전자 뉴트리노가 나타나지.

물리군 뮤온과 관계있는 뉴트리노도 있나요?

정교수 물론이야. 뮤온은 평균수명이 2.1969811마이크로초 정도로 아주 짧아.

물리군 마이크로초가 뭐예요?

정교수 1초의 십만분의 일을 1마이크로초라고 불러.

물리군 엄청 짧은 시간만 사는군요.

정교수 그렇지. 뮤온은 순식간에 붕괴해 다른 원소로 바뀐다네. 뮤온과 반뮤온의 붕괴 과정에서 뮤온 뉴트리노가 나와. 뮤온 뉴트리노는 1962년 세 명의 미국 과학자 레더먼(Leon M. Lederman), 슈워츠(Melvin Schwartz), 스타인버거(Jack Steinberger)가 발견했어. 뮤온 뉴트리노를 ν_μ, 뮤온 반뉴트리노를 $\overline{\nu_\mu}$라고 하면 뮤온과 반뮤온의 붕

괴 과정은 다음과 같아.

$$\mu^- \to e^- + \overline{\nu_e} + \nu_\mu$$

$$\mu^+ \to e^+ + \nu_e + \overline{\nu_\mu}$$

아이소스핀의 등장 _ 동일한 입자의 서로 다른 두 상태

정교수   전자의 스핀에 대해서는 알고 있나?

물리군   《반입자》책에서 읽었어요. 전자의 스핀은 S로 쓰는데

$$S = \frac{1}{2}$$

이고, 스핀의 z성분인 S_z는 $\frac{1}{2}$과 $-\frac{1}{2}$이 될 수 있어서 스핀에 따라 두 개의 양자 상태가 가능해요.

정교수   이제 스핀 개념을 양성자와 중성자에 도입하려고 해. 양성자와 중성자의 질량은 거의 비슷하지?

물리군   중성자가 조금 무겁네요.

정교수   1930년대 하이젠베르크와 위그너는 양성자와 중성자를 동일한 입자의 서로 다른 두 상태로 생각하기로 했어. 그리고 아이소스핀이라는 아이디어를 떠올렸지. 아이소스핀은 I로 쓰는데 이것의 z성분 I_z에 따라 서로 다른 상태가 나타난다고 보았어. 하이젠베르크는 각각의 아이소스핀 값에 따라 나타나는 I_z의 값이 다음과 같음을 알아

냈다네.

$$I = 0 \quad I_z = 0$$

$$I = \frac{1}{2} \quad I_z = \frac{1}{2} \quad I_z = -\frac{1}{2}$$

$$I = 1 \quad I_z = 1 \quad I_z = 0 \quad I_z = -1$$

$$I = \frac{3}{2} \quad I_z = \frac{3}{2} \quad I_z = \frac{1}{2} \quad I_z = -\frac{1}{2} \quad I_z = -\frac{3}{2}$$

그러니까 양성자와 중성자는 둘 다 $I = \frac{1}{2}$이지만, 양성자는 $I_z = \frac{1}{2}$이고 중성자는 $I_z = -\frac{1}{2}$이 되지.

물리군 재미있는 아이디어네요.

정교수 중요한 정의니까 잘 기억해 두게.

중간자 _ 핵력을 매개하는 입자가 필요해!

정교수 이번에는 중간자가 어떻게 발견되었는지 알아볼게.

물리군 중간자 아이디어는 일본 최초의 노벨상 수상자인 유카와 히데키가 낸 거죠?

정교수 맞아. 양성자와 중성자는 핵을 구성하는 입자이기 때문에 핵자라고 불러. 유카와 히데키는 핵자 사이의 힘에는 핵력을 매개하는 입자가 있어야 된다고 주장했지. 전자기 힘은 두 물체가 광자를 주고받는 과정이고, 핵력은 핵자들이 중간자를 주고받는 과정이라고 생각했어.

유카와 히데키의 예언대로 이러한 질량을 갖는 중간자가 발견되었다네. 영국의 실험물리학자 파월(Cecil Frank Powell)이 이것을 발견해 파이온이라고 불렀지. 파이온을 파이 중간자라고도 하는데 세 종류가 있어. 양의 전기를 띤 파이 중간자를 π^+, 음의 전기를 띤 파이 중간자를 π^-, 전기를 띠지 않은 파이 중간자를 π^0으로 나타내.

$\pi^+ = (+1, 139.6)$

$\pi^- = (-1, 139.6)$

$\pi^0 = (0, 135.0)$

물리군 파이온도 반입자가 있어요?

정교수 물론이야. π^+의 반입자는 π^-이고 π^0의 반입자는 $\overline{\pi^0}$로 써.

물리군 파이온도 붕괴하나요?

정교수 π^+나 π^-는 뮤온 또는 반뮤온으로 붕괴해. 식으로 나타내면 다음과 같아.

$\pi^+ \rightarrow \mu^+ + \nu_\mu$

$\pi^- \rightarrow \mu^- + \overline{\nu_\mu}$

이러한 붕괴에 걸리는 시간은 2.6×10^{-8}초인데 이게 바로 파이온의 수명이야.

π^0은 두 개의 광자로 붕괴돼.

$$\pi^0 \rightarrow 2\gamma$$

여기서 γ는 광자를 나타내지. 이 붕괴에 걸리는 시간은 48.4×10^{-17} 초인데 이것이 파이온의 수명이야.

파이온의 아이소스핀은 $I = 1$이야. 그러니까 세 종류의 파이온은 서로 다른 I_z값을 갖지.

$$\pi^+ : I_z = +1$$

$$\pi^0 : I_z = 0$$

$$\pi^- : I_z = -1$$

중간자가 발견된 후 과학자들은 물질을 구성하는 입자를 세 종류로 분류했어. 전자처럼 가벼운 입자를 경입자(lepton), 양성자처럼 무거운 입자를 중입자(baryon)라고 불러. 경입자와 중입자의 중간 질량을 가진 입자를 중간자(meson)라고 해.

경입자: 전자, 양전자, 전자 뉴트리노, 전자 반뉴트리노, 뮤온, 반뮤온, 뮤온 뉴트리노, 뮤온 반뉴트리노

중입자: 양성자, 중성자, 반양성자, 반중성자

중간자: $\pi^+, \pi^0, \pi^-, \overline{\pi^0}$

물리군 광자는 어디에 속해요?

정교수 광자는 물질을 구성하는 입자가 아니야. 전자기 힘을 매개하는 입자지. 그러니까 광자는 중입자도, 경입자도, 중간자도 아니야. 그리고 중간자와 중입자는 핵 안에서 작용하는 강력(핵력)과 관계있기 때문에 강입자(hadron)라고 불러.

이제 경입자수와 중입자수를 알아볼게. 경입자수는 L로 쓰는데 전자, 전자 뉴트리노, 뮤온, 뮤온 뉴트리노의 경입자수는 1로 정의해. 이들의 반입자인 양전자, 전자 반뉴트리노, 반뮤온, 뮤온 반뉴트리노의 경입자수는 -1로 정의하지. 중입자수는 B로 쓰는데 양성자, 중성자의 중입자수는 1로, 이들의 반입자인 반양성자, 반중성자의 중입자수는 -1로 정의해.

물리군 중간자수도 있나요?

정교수 그런 건 없어. 중간자는 경입자수도, 중입자수도 모두 0이야.

물리학자들은 모든 붕괴 반응에서 경입자수와 중입자수가 보존되는 것을 알아냈다. 예를 들어 다음 반응을 보자.

$$n \to p + e^- + \overline{\nu}_e$$

이 반응에서 붕괴 전의 중입자수는 1, 붕괴 후의 중입자수는 1이므로 중입자수는 보존된다. 붕괴 전의 경입자수는 0이고, 붕괴 후에는 전자의 경입자수가 1이고 전자 반뉴트리노의 경입자수가 -1이다. 즉, 전체 경입자수는 0이 되어 붕괴 전후의 경입자수도 보존된다.

이번에는 다음 반응을 보자.

$$\mu^- \to e^- + \overline{\nu}_e + \nu_\mu$$

이 반응에서 붕괴 전의 중입자수는 0, 붕괴 후의 중입자수는 0이므로 중입자수는 보존된다. 붕괴 전의 경입자수는 1이고 붕괴 후에는 전자의 경입자수가 1, 전자 반뉴트리노의 경입자수가 −1, 뮤온 뉴트리노의 경입자수가 1이다. 즉, 전체 경입자수는 1이 되어 붕괴 전후의 경입자수도 보존된다.

다음 반응도 살펴보자.

$$\pi^+ \to \mu^+ + \nu_\mu$$

이 반응에서 붕괴 전의 중입자수는 0, 붕괴 후의 중입자수는 0이므로 중입자수는 보존된다. 붕괴 전의 경입자수는 0이고, 붕괴 후에는 반뮤온의 경입자수가 −1이고 뮤온 뉴트리노의 경입자수가 1이다. 즉, 전체 경입자수는 0이 되어 붕괴 전후의 경입자수도 보존된다.

이번에는 다음 반응을 보자.

$$\pi^0 \to 2\gamma$$

이 반응은 붕괴 전후의 경입자수가 0으로 보존되고 중입자수도 0으로 보존된다.

타우 입자 _ 전자나 뮤온보다 무거운 경입자를 찾아서

정교수 이제 전자와 뮤온에 이어 세 번째로 발견된 타우 입자를 알아볼 거야. 우선 이것을 찾아낸 과학자 펄을 소개할게.

펄(Martin Lewis Perl, 1927~2014, 1995년 노벨 물리학상 수상)

펄은 미국 뉴욕주 뉴욕시에서 태어났다. 그의 부모는 러시아의 폴란드 지역에서 미국으로 이주한 유대인 이민자였다.

펄은 브루클린에 있는 브루클린 공과대학(현재 NYU-Tandon)에서 화학공학을 전공했다. 1948년에 학교를 졸업한 그는 제너럴 일렉트릭 컴퍼니에서 전자 진공관을 생산하는 공장의 화학 엔지니어로 일했다. 전자 진공관이 어떻게 작동하는지 배우기 위해 펄은 뉴욕 스케넥터디에 있는 유니온 칼리지의 원자물리학과 고급 미적분학 과정에 등록했다. 이를 계기로 물리학에 대한 관심이 커졌고 결국 1950년에 물리학과 대학원생이 되었다.

1955년 펄은 컬럼비아 대학에서 박사 학위를 받았다. 그의 논문 지도교수는 라비(I. I. Rabi)였다. 펄의 논문은 라비가 1944년 노벨 물리학상을 수상한 원자 빔 공명법을 사용해 나트륨의 핵 사중극자 모멘트를 측정하는 방법에 관한 연구였다. 박사 학위 취득 후 펄은 미시간 대학 교수가 되었다.

펄은 1963년 당시 캘리포니아에 건설 중이던 스탠퍼드 선형 가속기 센터(SLAC)로 자리를 옮겼다. 이때부터 전자나 뮤온보다 무거운 경입자를 찾는 문제에 매달렸다. 1974년부터 1977년까지 펄과 동료들은 SLAC와 로런스 버클리 국립연구소에서 새로운 경입자를 찾는 실험을 거듭했다. 결국 전자와 전하량은 같고 질량은 전자의 3477배로 무거운 새로운 경입자를 발견했다. 이 경입자는 τ로 나타내고 타우 입자로 불린다.

1997년 여름 DONUT(Direct Observation of the Nu Tau, E872) 팀은 타우 뉴트리노 ν_τ를 찾는 데 성공했다.

이렇게 해서 경입자족은 다음과 같이 3세대로 나뉘게 되었다.

제1세대 경입자: 전자, 양전자, 전자 뉴트리노, 전자 반뉴트리노
제2세대 경입자: 뮤온, 반뮤온, 뮤온 뉴트리노, 뮤온 반뉴트리노
제3세대 경입자: 타우 입자, 반타우 입자, 타우 뉴트리노, 타우 반뉴트리노

약력을 매개하는 입자 _질량이 0이 아닌 새로운 입자

물리군   전자기 힘은 광자를 주고받는 과정이고 핵력은 중간자를 주고받는 과정이잖아요? 그럼 약력은 어떤 입자를 주고받는 과정이죠?

정교수   약력은 베타 붕괴 과정에서 나타나는 힘이야. 다른 말로는 약한 상호작용이라고 하지. 1957년 슈윙거(Julian Schwinger)는 약력을 매개하는 입자가 양의 전기를 띤 W^+입자와 음의 전기를 띤 W^-입자와 전기를 띠지 않는 광자라고 주장했어.

물리군   광자는 전자기 힘에도 약력에도 관여하나요?

정교수   그렇지 않아. 슈윙거의 생각에 도전을 내민 세 명의 물리학자가 등장하지. 그들은 약력을 매개하는 입자 중 전기를 띠지 않은 입자는 광자가 아니라 질량이 0이 아닌 새로운 입자라고 보았어. 이 입자는 Z^0으로 불렸지. 이제 이 세 물리학자에 대해 간단히 알아볼게.

첫 번째로 소개할 과학자는 글래쇼다.

글래쇼는 1932년 미국 뉴욕에서 태어났다. 그의 아버지는 러시아 출신의 유대인 이민자였으며 배관공으로 일했다. 어릴 때부터 수학과 과학을 좋아했던 글래쇼는 과학 명문인 브롱크스 과학 고등학교에 입학했다. 1938년에 설립된 이 학교는 전 세계 중등학

글래쇼(Sheldon Lee Glashow, 1932~, 1979년 노벨 물리학상 수상)

교 중 과학 분야에서 가장 많은 노벨상 수상자를 배출했다. 현재까지 이 학교 출신의 노벨상 수상자는 9명이다.

1960년 브롱크스 과학 고등학교의 수업(출처: Jsclar/Wikimedia Commons)

글래쇼는 1954년 코넬 대학을 졸업하고 1959년 하버드 대학에서 노벨상 수상자인 슈윙거의 지도로 물리학 박사 학위를 받았다. 그 후 스탠퍼드 대학의 조교수로 일하다가 버클리 대학에서 1962년부터 1966년까지 부교수로 재직했다. 1966년에는 하버드 대학 물리학과 교수가 되었다.

두 번째로 소개할 물리학자는 와인버그다.

와인버그는 미국 뉴욕에서 태어났다. 그의 아버지는 유대인 이민자였으며 법원 속기사로 일했다. 16세 무렵 사촌이 물려준 화학 세트로 과학에 관

와인버그(Steven Weinberg, 1933~2021, 1979년 노벨 물리학상 수상, 사진 출처: Larry D. Moore/Wikimedia Commons)

심을 갖게 된 와인버그는 1950년에 브롱크스 과학 고등학교를 졸업했다. 그는 글래쇼와 같은 반이었다.

1954년 와인버그는 코넬 대학에서 학사 학위를 받고 프린스턴 대학으로 옮겨 1957년에 물리학 박사 학위를 받았다. 그 후 컬럼비아 대학(1957~1959)과 캘리포니아 대학(1959)에서 박사 후 연구원으로 일하다가 버클리 대학(1960~1966)에서 교수가 되었다.

세 번째로 소개할 물리학자는 살람이다.

살람(Mohammad Abdus Salam, 1926~1996, 1979년 노벨 물리학상 수상, 사진 출처: Molendijk, Bart/Anefo/Nationaal Archief)

살람은 1926년 1월 29일에 영국령 인도(현재 파키스탄)의 펀자브 지방에서 태어났다. 그는 라호르 대학 수학과를 졸업하고 영국으로 건너가 케임브리지의 세인트존스 칼리지에서 물리학을 공부했다.

케임브리지의 세인트존스 칼리지
(출처: DAVID ILIFF/Wikimedia Commons)

살람은 1951년 케임브리지의 캐번디시 연구소에서 이론물리학 박사 학위를 받았다. 그 후 그는 라호르의 Government College University의 수학과 교수 및 펀자브 대학의 수학과 교수가 되었다. 1953년 살람은 케임브리지로 돌아가 1954년에 세인트존스 칼리지의 수학 교수가 되었다. 1957년에는 임페리얼 칼리지 런던의 학장으로 초빙되었다.

물리군   세 명의 물리학자가 공동 연구를 했나요?

정교수   그렇지는 않아. 하지만 비슷한 시기에 독립적으로 같은 주제를 연구했어. 그 주제는 바로 약력과 전자기 힘의 통일이론이지. 세 사람은 약력과 전자기 힘의 통일이론을 위해서는 Z^0 입자가 반드시 필요하다는 것을 알아냈다네. 그들은 독립적으로 약력과 전자기 힘의 통일이론에 대한 연구 결과를 발표했어. 이 업적으로 1979년 노벨 물리학상을 공동 수상했지.

베타 붕괴 과정에서 W^+입자와 W^-입자가 나타나는 과정은 다음과 같다. 우선 중성자는 양성자와 W^-입자로 붕괴된다.

$$n \rightarrow p + W^-$$

그리고 짧은 시간이 경과한 후에 W^-입자는 다음과 같이 붕괴된다.

$$W^- \rightarrow e^- + \overline{\nu_e}$$

마찬가지로 양성자가 중성자로 붕괴되는 과정은

$$p \rightarrow n + W^+$$

이고,

$$W^+ \rightarrow e^+ + \nu_e$$

이다.

이렇게 전하가 바뀌는 경우의 약한 상호작용은 W^+입자와 W^-입자로 설명된다. 세 사람은 전하가 바뀌지 않는 약한 상호작용은 Z^0입자로 설명할 수 있다고 생각했다.

약력을 매개하는 입자의 발견 _ 양성자와 반양성자를 충돌시켜

물리군 W^+입자, W^-입자, Z^0입자는 발견되었나요?

정교수   물론이지. 이 입자들을 발견한 사람은 이탈리아의 루비아야.

루비아(Carlo Rubbia, 1934~,
1984년 노벨 물리학상 수상.
사진 출처: Markus Pössel/Wikimedia Commons)

루비아는 1934년 슬로베니아 국경에 있는 이탈리아 마을 고리치아에서 태어났다. 그는 피사 대학에서 물리학을 공부했고 1957년에 박사 학위를 받았다. 학위 취득 후 그는 박사 후 연구를 위해 미국으로 건너갔다. 그리고 약 1년 반 동안 컬럼비아 대학에서 뮤온 붕괴와 핵포획 실험을 수행했다.

새로 설립된 CERN에 합류하기 전에 루비아는 로마 대학에서 일하러 유럽으로 돌아갔다. 거기서 약한 상호작용의 구조에 대한 실험에 참여했다. CERN은 서로 충돌하는 양성자의 역회전 빔을 사용하는 새로운 유형의 가속기인 Intersecting Storage Rings(ISR)를 시운전했다. 루비아와 동료들은 그곳에서 실험을 진행하며 약력을 연구했다.

ISR(출처: CERN)

1976년 루비아는 CERN의 슈퍼 프로톤 싱크로트론(SPS)을 개조해 양성자와 반양성자를 충돌시키는 양성자–반양성자 충돌기를 제안했다. 이 충돌기는 1981년부터 가동했으며, 1983년 초에 루비아가 이끄는 팀이 약력의 매개 입자인 W^+입자, W^-입자, Z^0입자를 발견했다. 이 발견으로 루비아는 1984년 노벨 물리학상을 수상했다.

SPS(출처: Gillis/Wikimedia Commons)

Z^0입자를 발견한 거품상자(출처: Fanny Schertzer/Wikimedia Commons)

　루비아가 발견한 세 가지 매개 입자의 전하와 정지에너지는 다음과 같다.

$W^+ = (+1, 80.4\text{GeV})$

$W^- = (-1, 80.4\text{GeV})$

$Z^0 = (0, 91.2\text{GeV})$

네 번째 만남

기묘한 입자의 출현

케이온을 발견한 두 과학자 _파이온 다음으로 발견된 중간자

물리군 중간자는 파이온뿐인가요?

정교수 그렇지 않아. 파이온 외에 많은 중간자가 발견되었어. 파이온 다음으로 발견된 중간자는 케이온이야. 케이온을 찾아낸 두 과학자를 소개할게. 먼저 로체스터(George Dixon Rochester, 1908~2001)에 대해 알아볼까?

로체스터는 영국 월센드에서 태어났다. 지역 초등학교를 다닌 그는 1920년에 월센드 문법학교에 입학하여 화학과 물리학에서 좋은 성적을 거두었다. 뉴캐슬의 암스트롱 칼리지에서는 장학금을 받았다. 그는 1930년 커티스(W. E. Curtis)의 지도하에 물리학에서 일등으로 졸업했고, 대학원 장학금을 받고 1931년에 커티스의 연구 그룹에 합류했다.

1934년부터 1935년까지 로체스터는 스톡홀름 대학 물리학 연구소의 에릭 훌텐(Erik Hulthén) 교수와 함께 밴드 스펙트럼을 연구했다. 암스트롱 대학에 재학 중이던 그는 1932년에 석사 학위를, 1937년에 박사 학위를 취득했다.

1937년 로체스터는 맨체스터 빅토리아 대학의 조교수로 임명되어 우주선(cosmic rays)을 연구했다. 1939년에 전쟁이 발발하자 스카버러 근처의 스택스턴에서 새롭게 운영된 레이더 기지에서 근무했다. 그는 우주선 물리학자인 야노시(Lajos Jánossy)와 함께 해수면에서 우주선의 양을 연구했고, 이 연구는 전쟁 후에도 계속되었다.

1955년부터 로체스터는 더럼 대학의 물리학과 교수로 재직했다. 그는 노벨 물리학상 후보에 22번 지명되었지만 상을 받는 데는 실패했다.

두 번째로 소개할 물리학자는 버틀러(Clifford Charles Butler, 1922~1999)이다.

버틀러는 1922년 5월 20일 영국 레딩에서 태어났다. 그는 레딩 스쿨과 레딩 대학에 다녔으며, 그곳에서 과학 학사 학위와 철학 박사 학위를 받았다. 1945년 그는 맨체스터 대학의 물리학 조교수로 임명되었다.

1953년 버틀러는 맨체스터를 떠나 임페리얼 칼리지 런던에서 고에너지 핵물리학 그룹을 이끌었다. 1970년 임페리얼 칼리지에서 사임한 그는 교육에 영향력 있는 자선 재단인 너필드 재단(Nuffield Foundation)의 이사직을 수락했다. 그곳에서 그는 고등교육의 연구 및 혁신을 위한 그룹, 법률과 사회를 위한 프로그램, 레딩 대학의 농업 전략 센터를 설립했다. 1975년에는 러프버러 대학의 부총장으로 임명되어 1985년에 은퇴할 때까지 그 직을 유지했다.

V입자 _ 성차별을 극복하고

정교수 케이온은 처음에 V입자라고 불렸어.

물리군 왜죠?

정교수 1947년에 조지 로체스터와 클리퍼드 버틀러가 우주선 관측 중 발견했는데, 윌슨의 안개상자 안에 V 모양의 비적을 남겼기 때문이지.

V입자는 양의 전기를 띤 것과 음의 전기를 띤 것, 전기를 띠지 않은 것의 세 종류가 발견되었다. 이 중 전기를 띠는 입자는 두 개의 파이온으로 붕괴했다. 예를 들어 양의 전기를 띠는 V입자는 양의 전기를 띠는 파이온과 중성의 파이온으로 붕괴했다. 과학자들은 이렇게 두 개의 파이온으로 붕괴하는 V입자를 세타 입자라고 불렀다. 이 중 양의 전기를 띤 세타 입자는 θ^+라고 쓴다. 즉, 세타 입자의 붕괴 식은 다음과 같다.

$$\theta^+ \rightarrow \pi^+ + \pi^0$$

1949년 브리스틀 대학에서 세실 파월(Cecil Powell)의 연구생 로즈메리 브라운(Rosemary Brown)은 세타 입자와 질량이 거의 같은 입자가 3개의 파이온으로 붕괴하면서 'k'자 모양의 궤적을 만드는 것을 발견했다. 과학자들은 이 입자를 타우 입자[1]로 불렀다.

---

1) 앞에서 등장한 타우 입자와는 다르다. 이 이름은 얼마 후 사라진다.

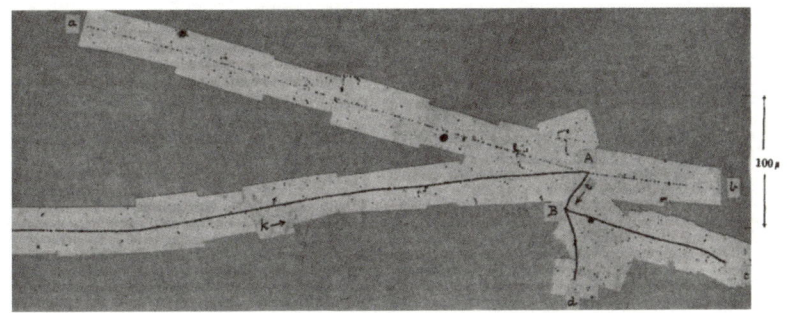

케이온의 붕괴

예를 들어 양의 전기를 띤 타우 입자를 τ^+로 쓰면 붕괴 식은

$$\tau^+ \to \pi^+ + \pi^+ + \pi^-$$

이다. 두 입자는 질량, 스핀, 수명이 모두 같았다.

물리군 세타 입자와 타우 입자는 같은 입자인가요?

정교수 많은 과학자는 두 입자가 다른 입자라고 생각했어.

물리군 어째서요?

정교수 두 입자의 패리티가 다르기 때문이지.

물리군 패리티가 뭐예요?

정교수 입자는 파동함수로 표현할 수 있어. 예를 들어 어떤 입자의 파동함수가 $\psi(x)$라고 해봐. 이때 x를 $-x$로 바꾸는 연산을 패리티 연산이라 하고 P로 나타내. 즉,

$$P\psi(x) = \psi(-x)$$

가 되지. 여기서 패리티 연산자를 두 번 취하면

$$P^2 \psi(x) = \psi(x)$$

이므로

$$P^2 = I$$

인 항등연산이 돼. 그러니까 파동함수를 패리티 연산자에 의해 두 종류로 구분할 수 있어.

$$P\psi(x) = +\psi(x) \quad \text{(양의 패리티를 가질 때)}$$

$$P\psi(x) = -\psi(x) \quad \text{(음의 패리티를 가질 때)}$$

세타 입자의 붕괴를 살펴볼까? 붕괴 전은 θ^+의 파동함수 ψ_{θ^+}에 의해 묘사되고, 붕괴 후에는 $\psi_{\pi^+}\psi_{\pi^0}$에 의해 묘사되지. 이 경우 붕괴 후의 패리티가 +1이 돼.

물리군 왜 패리티가 +1인가요?

정교수 음수와 음수의 곱이 양수이기 때문이야. 즉,

$$P(\psi_{\pi^+}\psi_{\pi^0}) = \psi_{\pi^+}(-x)\psi_{\pi^0}(-x) = (-\psi_{\pi^+}(x))(-\psi_{\pi^0}(x)) = \psi_{\pi^+}\psi_{\pi^0}$$

이거든. 하지만 타우 입자의 붕괴를 봐. 붕괴 전은 τ^+의 파동함수 ψ_{τ^+}에 의해 묘사되고, 붕괴 후에는 $\psi_{\pi^+}\psi_{\pi^+}\psi_{\pi^-}$에 의해 묘사돼. 이 경우

$$P(\psi_{\pi^+}\psi_{\pi^+}\psi_{\pi^-}) = \psi_{\pi^+}(-x)\psi_{\pi^+}(-x)\psi_{\pi^-}(-x)$$

$$= (-\psi_{\pi^+}(x))(-\psi_{\pi^+}(x))(-\psi_{\pi^-}(x))$$

$$= -\psi_{\pi^+}\psi_{\pi^+}\psi_{\pi^-}$$

이므로 붕괴 후 패리티가 -1이어야 해. 그러니까 붕괴 전후에 패리티가 보존되어야 한다면 세타 입자와 타우 입자는 서로 다른 입자지. 하지만 두 입자는 질량, 스핀과 같은 모든 성질이 같았어. 과학자들은 여전히 이 두 입자가 같은 게 아닐까 궁금했지. 이것을 타우-세타 미스터리라고 불러. 이 문제를 해결한 과학자들은 중국계 미국인 삼총사야.

세 사람 중 먼저 리정다오에 대해 알아볼게.

리정다오(李政道, 1926~2024, 1957년 노벨 물리학상 수상)

리정다오는 중국 상하이에서 태어났다. 그의 아버지는 난징 대학의 첫 번째 졸업생 중 한 명으로, 중국의 초기 현대식 합성 비료 개발에 참여한 화학 산업가이자 상인이었다.

리정다오는 상하이 쑤저우 대학 부속고등학교에서 중등교육을 받았다. 제2차 중일전쟁으로 학업이 중단되어 그는 졸업장을 받지 못했다. 그럼에도 1943년 저장 대학 화학공학과에 입학했다. 그는 물리학에 재능을 발견하면서 저장 대학 물리학과로 편입하여 1943년부터 1944년까지 공부했다.

1946년 리정다오는 미국 시카고 대학에 진학해 페르미(Enrico Fermi) 교수의 박사 과정생이 되었다. 그리고 1950년 〈백색왜성의 수소 함량〉 연구로 박사 학위를 받았다. 그는 1953년에 컬럼비아 대학 교수가 되었고, 주로 양자장 이론의 해결 가능한 모델을 연구했다.

이번에는 양전닝에 대해 살펴보자.

양전닝(楊振寧, 1922~2025, 1957년 노벨 물리학상 수상)

양전닝은 중국 안후이성 허페이에서 태어났다. 그의 아버지는 수학자였다. 양전닝은 베이징에서 초등학교와 고등학교를 다녔다. 1937년 가을, 일본군이 중국을 침략하자 그는 가족과 함께 허페이로 이사했다.

양전닝은 1942년에 국립 남서부 연합 대학을 졸업하고 대학원에서 통계역학을 연구했다. 1944년 그는 중일전쟁(1937~1945) 동안 쿤밍으로 이전한 칭화 대학에서 이학 석사 학위를 받았다.

1946년 1월, 양전닝은 미국 시카고 대학에 입학해 에드워드 텔러(Edward Teller)의 지도를 받아 1948년에 박사 학위를 받았다. 양전닝은 페르미(Enrico Fermi)의 조교로 1년간 시카고 대학에 머물렀다. 1949년 그는 뉴저지주의 프린스턴 고등연구소에 초빙되어, 그곳에서 리정다오와 공동 연구를 하기 시작했다. 그는 1955년 프린스턴 대학의 교수를 거쳐 1965년 뉴욕 대학(스토니브룩)의 물리학과 교수가 되어 새로 설립된 이론물리학 연구소의 초대 소장이 되었다.

마지막으로 소개할 물리학자는 우젠슝이다.

우젠슝(吳健雄, 1912~1997)

우젠슝은 중국 장쑤성 타이창시 류허에서 태어났다. 그녀의 어머니는 교사였으며, 그녀의 아버지는 엔지니어이자 사회적 진보주의자였다. 그는 1913년 제2혁명에 참여했고, 혁명이 실패한 후 류허로 이사해 밍더 중학교를 세웠다. 우젠슝은 밍더 중학교에서 교육을 받았다. 그녀는 다른 아이들처럼 밖에서 노는 대신 새로 발명된 라디오를 들으며 지식을 얻는 것을 좋아했다. 또한 시와 논어 같은 중국 고전이나 민주주의에 관한 서양 문학을 즐겨 읽었다.

밍더 중학교(출처: 白色瑰宝/Wikimedia Commons)

우젠슝은 1923년 고향을 떠나 집에서 50마일(80킬로미터) 떨어진 쑤저우 제2 여자 사범학교에 진학했다. 이 학교는 교사 양성과 일반 고등학교 과정이 있는 기숙 학교였다. 우젠슝은 이곳에서 과학에 관심을 가지게 되었다.

1929년 학교를 수석으로 졸업한 우젠슝은 난징에 있는 국립중앙대학에 입학했다. 그녀는 처음에 수학을 전공했지만 나중에 물리학으로 전공을 바꾸었다. 졸업 후 2년 동안은 저장 대학에서 조교로 일했다. 그녀의 지도교수는 미시간 대학에서 박사 학위를 취득한 여성 교수 구징웨이였다.

우젠슝은 미국 유학을 결심했다. 그녀는 미시간 대학에서 여성이 정문으로 다니는 것조차 허용되지 않는 사실을 알게 되었다. 미국 사회의 성차별에 충격받은 그녀는 더 진보적인 캘리포니아 대학 버클리에서 공부하기로 결정했다.

친구와 함께 우젠슝
(출처: Knottinghill/Wikimedia Commons)

버클리에서 우젠슝은 사이클로트론의 발명자 로런스 교수와 이탈리아 출신의 물리학자 세그레 교수에게 실험을 배웠다. 그녀와 세그레는 제논을 포함한 베타 붕괴 연구를 수행했다. 우젠슝은 1940년 6월에 박사 학위를 취득하고 방사선 연구소에서 박사 후 연구원으로 일했다.

1944년 3월, 우젠슝은 컬럼비아 대학의 맨해튼 프로젝트의 대체 합금 재료(SAM) 연구소에 합류했다. 그녀는 이곳에서 우라늄 농축을 위한 기체 확산 연구를 했고, 방사선 검출기를 개발하는 임무를 맡았다.

1945년 8월, 전쟁이 끝난 후 우젠슝은 컬럼비아 대학의 교수가 되었다. 이때 그녀의 주 관심사는 페르미가 제안한 베타 붕괴 연구였다.

물리군 세 명의 중국인 과학자는 어떻게 타우-세타 미스터리를 풀었나요?

정교수 리정다오와 양전닝은 시카고 대학에서 만났어. 두 사람의 첫 공동 논문은 1949년에 발표되었지. 그들은 타우 입자와 세타 입자가 질량, 스핀, 수명에서 같은 값을 가지므로 두 입자는 틀림없이 같은 입자이어야 한다고 생각했어.

두 사람은 하나의 입자가 패리티가 보존되는 붕괴를 하거나 패리티 보존이 깨지는 붕괴를 할 수 있다면 타우 입자와 세타 입자가 같은 입자여도 된다는 놀라운 생각을 떠올렸어. 그 전까지 패리티 보존은 필수였는데, 리정다오와 양전닝은 붕괴에서 패리티 보존이 필수가 아니라고 주장한 거지. 세타와 타우 문제뿐만 아니라 일반적인 약한 상호작용(베타 붕괴나 역 베타 붕괴 같은 과정)에서 패리티 보존이 깨질 수 있다는 내용을 1956년에 논문으로 발표했다네. 이 논문에는 구체적인 실험 시나리오도 적혀 있었어.

물리군 패리티가 보존되지 않는 반응을 실험에서 찾을 수 있으면 리정다오와 양전닝의 생각이 옳은 거군요.

정교수 맞아. 이 일은 실험물리학자가 해결해야 했지. 이 실험에 뛰어든 사람이 바로 우젠슝이야. 그녀는 리정다오와 양전닝의 논문에서 묘사된 코발트 60의 붕괴 과정을 실험해 보기로 했지. 1956년 5월 말에 본격적으로 작업을 시작한 그녀는 남편과의 여행도 취소하고 실험에 매달렸어. 그해 12월에 그녀는 코발트 60의 붕괴에서 나오는 베타 붕괴 과정에서 패리티가 보존되지 않는다는 실험 결과를 얻었다네. 우젠슝의 실험으로 리정다오와 양전닝은 이듬해 노벨 물리학상을 수상했지만 우젠슝은 아쉽게도 노벨상을 수상하지 못했지.

케이온 _ 기묘한 입자

정교수 이렇게 타우 입자와 세타 입자가 같은 입자라는 것이 밝혀지고, 이 중간자의 이름은 케이온으로 명명되었어. 그리고 세 종류의 케이온을 K^+, K^0, K^- 로 쓰기로 했다네. 그러니까 K^+의 반입자는 K^-이고, K^0의 반입자는 $\overline{K^0}$이지.

세 종류의 케이온의 질량은 다음과 같아.

$K^+ = (+1, 493.7)$

$K^- = (-1, 493.7)$

$K^0 = (0, 497.6)$

물리군 K^0과 $\overline{K^0}$는 어떻게 붕괴하죠?

정교수 K^0과 $\overline{K^0}$ 각각이 어떻게 붕괴하는지, 질량이 얼마인지는 알 수 없어. 다만 이들을 나타내는 두 종류의 파동함수가 중첩된 형태로 활동하지. 수명이 짧은 경우의 파동함수를 ψ_S, 수명이 긴 경우의 파동함수를 ψ_L이라고 하면

$$\psi_S = \frac{1}{\sqrt{2}}(\psi_{K^0} - \psi_{\overline{K^0}})$$

$$\psi_L = \frac{1}{\sqrt{2}}(\psi_{K^0} + \psi_{\overline{K^0}})$$

가 돼. 우리가 실험으로 발견하는 입자의 상태는 K^0과 $\overline{K^0}$가 아니라 이들의 파동함수가 중첩된 형태로 나타나. 파동함수 ψ_S에 대응하는 입자를 K_S라 하고, 파동함수 ψ_L에 대응하는 입자를 K_L이라고 부른다네. K_S의 붕괴 식은 다음과 같아.

$$K_S \rightarrow \pi^+ + \pi^-$$

또는

$$K_S \to \pi^0 + \pi^0$$

그리고 K_L의 붕괴 식은 다음과 같아.

$$K_L \to \pi^0 + \pi^0 + \pi^0$$

또는

$$K_L \to \pi^+ + \pi^0 + \pi^-$$

K_S의 평균수명은 8.95×10^{-11}(초)이고 K_L의 평균수명은 5.12×10^{-8}(초)야. 따라서 K^0과 $\overline{K^0}$가 혼합된 빔이 발생하면 이 빔은 K_S와 K_L이 혼합된 빔이니까 K_S가 먼저 붕괴되고 나중에는 K_L만 남지.

물리군 케이온의 아이소스핀은 1인가요?

정교수 케이온의 아이소스핀은 $I = \frac{1}{2}$ 이야. 그러니까 허용 가능한 상태는

$$I_z = \frac{1}{2}, -\frac{1}{2}$$

의 두 가지라네.

물리군 케이온은 세 종류잖아요?

정교수 하나의 허용 가능한 상태에 두 종류의 케이온이 존재해. 즉,

$K^+, \overline{K^0} : I = 0, I_z = \dfrac{1}{2}$ 인 상태

$K^-, K^0 : I = 0, I_z = -\dfrac{1}{2}$ 인 상태

이지.

물리군 케이온은 참 기묘한 입자네요.

정교수 맞아. 그래서 물리학자들은 입자의 기묘한 정도를 나타내기 위해 새로운 보존량인 기묘도(strangeness)를 도입했어.

기묘도 _ 입자의 속성을 나타내는 새로운 수치

정교수 기묘도는 일본의 니시지마와 미국의 겔만이 도입했지. 겔만에 대해서는 나중에 얘기하고 먼저 니시지마가 어떤 사람인지 알아볼게.

니시지마 가즈히코(西島和彦, 1926~2009.
사진 출처: 日本学士院)

니시지마는 1926년 10월 4일 일본 쓰치우라에서 태어났다. 그는 1948년 도쿄 대학에서 물리학 학사 학위를 받았고, 1955년 오사카 대학에서 핵력의 퍼텐셜에 대한 논문으로 박사 학위를 받았다.

1950년 오사카 대학에 재학 중이던 니시지마는 난부 요이치로 밑에서 강한 상호작용과 V입자 이론을 연구했다. 1956년부터 1958년까지 니시지마는 하이젠베르크의 초청으로 독일 괴팅겐 대학에서 일했다. 그는 1958년 미국으로 건너가 프린스턴 고등연구소에서 일했다. 1년 반 후에는 일리노이 대학 어배너-샘페인 캠퍼스의 교수가 되었다. 1966년 그는 도쿄 대학으로 돌아와 이론물리학 연구 그룹을 설립하고 물리학과 교수로 일했다.

물리군 기묘도가 뭐예요?

정교수 기묘도는 S로 나타내는데 기묘하지 않은 입자들의 기묘도는 0이야. 지금까지 나온 입자 중에서 양성자, 중성자, 파이온 등은 기묘도가 0이지. 하지만 케이온은 기묘한 입자라네. 니시지마와 겔만은 이 입자의 기묘도를 다음과 같이 정의했어.

$K^+, K^0 : S = 1$

$K^-, \overline{K^0} : S = -1$

물리군 기묘도가 음수일 수도 있군요!

정교수 그렇지. 기묘도는 정수로 나타내. 그러니까 입자의 속성은 전하, 아이소스핀, 기묘도, 패리티(P)에 의해 묘사되지. 앞으로 전하는 Q로 쓸게. 전자의 경우 $Q = -1$이고 양성자의 경우 $Q = +1$이야. 이 정수들이 입자 연구에 아주 중요해.

물리군 케이온 말고 또 다른 중간자도 있나요?

정교수 입자가속기 덕분에 많은 중간자가 발견되었어.

1961년 페브스너(Aihud Pevsner)는 베바트론에서 파이온-핵자 충돌로 생긴 새로운 중간자인 에타 입자를 발견했다. 에타 입자는 η로 나타낸다.

$$\eta = (0, 547.9)$$

이며 평균수명은

$$5 \times 10^{-19} (초)$$

로 다음과 같이 붕괴한다.

$$\eta \rightarrow \gamma + \gamma$$

또는

$$\eta \rightarrow \pi^0 + \pi^0 + \pi^0$$

또는

$$\eta \rightarrow \pi^+ + \pi^0 + \pi^-$$

에타 입자의 전하, 아이소스핀, 기묘도, 패리티는 다음과 같다.

$Q = 0$

$I = 0$

$S = 0$

$P = -1$

새로운 중입자 _ 람다, 시그마, 크시, 델타

물리군 양성자와 중성자 말고 또 다른 중입자들이 발견되었나요?

정교수 물론이야. 입자가속기 덕분에 많은 새로운 중입자가 발견되었지.

1950년 10월, 호주 멜버른 대학의 호퍼(V. D. Hopper)와 비스와스(S. Biswas)는 열기구를 타고 7만 피트(21,000미터) 상공에서 우주선을 관측해 람다 입자라는 중입자를 발견했다. 이 입자는 Λ로 표기하며

$\Lambda = (0, 1115.7)$

이다. 이 역시 기묘한 입자로 분류되는데 아이소스핀, 기묘도는 다음과 같다.

$I = 0$

$S = -1$

또 다른 중입자로 새로운 기묘한 입자인 시그마 입자가 발견되었다. 시그마 입자는 Σ로 나타내는데, 전하에 따라 다음과 같이 세 종류로 구분한다.

$\Sigma^+ = (+1, 1189.37)$

$\Sigma^0 = (0, 1192.64)$

$\Sigma^- = (-1, 1197.45)$

시그마 입자는 아이소스핀이 $I = 1$인 입자이며, 아이소스핀의 z성분과 기묘도는 다음과 같다.

$\Sigma^+ : I_z = +1 \quad S = -1$

$\Sigma^0 : I_z = 0 \quad S = -1$

$\Sigma^- : I_z = -1 \quad S = -1$

Σ^+의 붕괴 과정은 다음과 같다.

$$\Sigma^+ \to p + \pi^0$$

또는

$$\Sigma^+ \to n + \pi^+$$

Σ^0과 Σ^-의 붕괴 과정은 다음과 같다.

$$\Sigma^0 \to \Lambda + \gamma$$

$$\Sigma^- \to n + \pi^-$$

또 다른 중입자로 크시 입자가 발견되었다. 크시 입자는 Ξ로 표시한다. 음의 전기를 띤 Ξ^-가 1952년 맨체스터 그룹의 우주선 실험에서, 전기를 띠지 않는 Ξ^0은 1959년 로런스 버클리 연구소에서 발견되었다.

$$\Xi^0 = (0, 1314.86)$$

$$\Xi^- = (-1, 1321.71)$$

크시 입자는 아이소스핀이 $I = \frac{1}{2}$인 입자이며, 아이소스핀의 z성분과 기묘도는 다음과 같다.

$$\Xi^0 : I_z = +\frac{1}{2} \quad S = -2$$

$$\Xi^- : I_z = -\frac{1}{2} \quad S = -2$$

두 크시 입자의 붕괴와 평균수명은 다음과 같다.

$\Xi^0 \rightarrow \Lambda + \pi^0$ (2.9×10^{-10}초)

$\Xi^- \rightarrow \Lambda + \pi^-$ (1.6×10^{-10}초)

1950년대 중반 시카고 대학의 사이클로트론과 카네기 공과대학에서 싱크로-사이클로트론 실험을 통해 새로운 중입자인 델타 입자가 발견되었다. 이 입자는 Δ로 나타내는데 모두 네 종류이다.

$\Delta^{++} = (+2, 1232)$

$\Delta^+ = (+1, 1232)$

$\Delta^0 = (0, 1232)$

$\Delta^- = (-1, 1232)$

델타 입자는 아이소스핀이 $I = \frac{3}{2}$인 입자이며, 아이소스핀의 z성분과 기묘도는 다음과 같다.

$\Delta^{++}: I_z = +\frac{3}{2} \quad S = 0$

$\Delta^+: I_z = +\frac{1}{2} \quad S = 0$

$\Delta^0: I_z = -\frac{1}{2} \quad S = 0$

$$\Delta^- : I_z = -\frac{3}{2} \quad S = 0$$

네 델타 입자의 붕괴와 평균수명은 다음과 같다.

$$\Delta^{++} \to p + \pi^+ \ (5.63 \times 10^{-24}초)$$

$$\Delta^+ \to \pi^+ + n \ 또는 \ \pi^0 + p \ (5.63 \times 10^{-24}초)$$

$$\Delta^0 \to \pi^0 + n \ 또는 \ \pi^- + p \ (5.63 \times 10^{-24}초)$$

$$\Delta^- \to \pi^- + n \ (5.63 \times 10^{-24}초)$$

니시지마-나카노-겔만 법칙 _세 사람이 발견한 재미있는 공식

정교수 이번에는 니시지마, 나카노, 겔만이 발견한 재미있는 공식을 알려줄게. 이것은 1953년에 나카노와 니시지마가 처음 발견했고, 겔만도 1956년에 독립적으로 발견했어.

물리군 어떤 공식인가요?

정교수 입자의 아이소스핀의 z성분(I_z)과 전하(Q), 중입자수(B), 기묘도(S) 사이의 관계식으로 다음과 같아.

$$Q = I_z + \frac{1}{2}(B + S)$$

예를 들어 양성자를 보자. 양성자의 경우

$$Q = 1$$

이고

$$I_z + \frac{1}{2}(B+S) = \frac{1}{2} + \frac{1}{2}(1+0) = 1$$

이 되어 공식을 만족한다.

중성자는

$$Q = 0$$

이고

$$I_z + \frac{1}{2}(B+S) = -\frac{1}{2} + \frac{1}{2}(1+0) = 0$$

이 되어 공식을 만족한다.

K^+입자는

$$Q = 1$$

이고

$$I_z + \frac{1}{2}(B+S) = \frac{1}{2} + \frac{1}{2}(0+1) = 1$$

이 되어 공식을 만족한다.

에타 입자는

$$Q = 0$$

이고

$$I_z + \frac{1}{2}(B+S) = 0 + \frac{1}{2}(0+0) = 0$$

이 되어 공식을 만족한다.

람다 입자는

$$Q = 0$$

이고

$$I_z + \frac{1}{2}(B+S) = 0 + \frac{1}{2}\{1+(-1)\} = 0$$

이 되어 공식을 만족한다.

Ξ^- 입자는

$$Q = -1$$

이고

$$I_z + \frac{1}{2}(B+S) = -\frac{1}{2} + \frac{1}{2}\{1+(-2)\} = -1$$

이 되어 공식을 만족한다.

Δ^{++}입자는

$$Q = 2$$

이고

$$I_z + \frac{1}{2}(B+S) = \frac{3}{2} + \frac{1}{2}(1+0) = 2$$

가 되어 공식을 만족한다.

여기서

$$Y = B + S$$

를 초전하(hypercharge)라고 부른다. 그러므로 니시지마-나카노-겔만 공식은

$$Q = \frac{1}{2}Y + I_z$$

로 쓸 수 있다.

다섯 번째 만남
•
쿼크모형

사카타 모형 _ 페르미-양 모형을 모든 강입자로 확장하다

정교수 이제 쿼크모형이 나오는 데 큰 도움을 준 사카타 모형에 대해 이야기해 볼게.

사카타 쇼이치(坂田昌一, 1911~1970)

사카타 쇼이치는 1911년 1월 18일 일본 도쿄에서 태어났다. 그의 아버지는 사카타의 대부가 된 가쓰라 다로 총리의 비서였다. 1924년 효고현의 고난 중학교에 다니던 사카타는 물리학자 아라카츠 분사쿠에게 가르침을 받았다. 1926년부터 1929년까지 고난 고등학교에 재학하던 시절, 사카타는 유명한 물리학자 이시와라 준의 강의를 들었다.

사카타는 1930년에 교토 대학에 입학했다. 대학 2학년 때 그는 니시나 요시오로부터 양자역학을 배웠다. 이때 유카와 히데키와 도모나가 신이치로를 알게 되었다. 대학을 졸업한 후, 사카타는 1933년에 이화학 연구소(RIKEN)에서 도모나가 신이치로, 니시나 요시오와 함께 일했다. 1934년에는 오사카 대학으로 옮겨 유카와 히데키와 일했다. 유카와 히데키는 1935년에 중간자 이론에 관한 첫 번째 논문을 발표했다. 사카타는 중간자 이론의 발전을 위해 그와 공동 연구를 했다.

교토 대학(출처: Soraie8288/Wikimedia Commons)

사카타는 1942년 10월 나고야 대학의 교수로 부임하여 사망할 때까지 재직했다. 나고야로 돌아온 후, 사카타와 그의 나고야 그룹은 니시지마–겔만 공식 뒤에 숨겨진 물리학을 밝히는 연구를 시작했다. 1956년에 사카타는 세 개의 기본 입자가 모든 강입자를 구성한다는 사카타 모형을 발표했다.

물리군 사카타 모형이 뭐죠?

정교수 사카타 모형이 나오기 전에 페르미-양 모형이 있었어. 페르미와 양전닝은 파이온이 양성자, 중성자와 이들의 반입자의 복합 입자라고 생각했지. 예를 들면 다음과 같아.

$$\pi^+ = p\bar{n}$$

$$\pi^- = n\bar{p}$$

물리군 이상해요. 파이온의 질량이 양성자와 반양성자의 질량의 합보다 작은데, 어떻게 파이온이 양성자와 반양성자의 복합 입자가 될 수 있나요?

정교수 페르미와 양전닝이 생각한 식은

$$p + \bar{n} \to \pi^+ + X$$

였어. 여기서 X는 에너지를 가진 미지의 입자야. 입자들의 세계에서 질량은 보존될 필요가 없다네. 다만 에너지가 보존되지.

사카타는 케이온이 발견된 후 페르미-양 모형을 모든 강입자로 확장했다. 그는 모든 강입자가 세 개의 기본 입자와 그 입자들의 반입자로 이루어져 있다고 생각했다. 사카타는 세 개의 기본 입자로 양성자(p)와 중성자(n), 람다 입자(Λ)를 선택했다. 이 세 입자의 아이소스핀과 기묘도는 다음과 같다.

	p	n	Λ
I_z	$\dfrac{1}{2}$	$-\dfrac{1}{2}$	0
S	0	0	-1

이 입자들의 반입자에 대한 아이소스핀과 기묘도는 다음과 같다.

	$\bar{p}$	$\bar{n}$	$\bar{\Lambda}$
I_z	$-\dfrac{1}{2}$	$\dfrac{1}{2}$	0
S	0	0	1

예를 들어 케이온의 경우는 다음과 같이 구성된다.

$K^+ = p\bar{\Lambda}$

$K^0 = n\bar{\Lambda}$

$K^- = \Lambda\bar{p}$

$\overline{K^0} = \Lambda\bar{n}$

첫 줄에서

$$I_z(K^+) = \frac{1}{2}$$

이고,

$$I_z(p) = \frac{1}{2}, \ I_z(\overline{\Lambda}) = 0$$

이 되어,

$$I_z(p\overline{\Lambda}) = \frac{1}{2} + 0 = \frac{1}{2}$$

이므로 아이소스핀의 z성분이 보존된다. 이번에는 기묘도를 확인하자.

$$S(K^+) = 1$$

이고

$$S(p\overline{\Lambda}) = S(p) + S(\overline{\Lambda}) = 0 + 1 = 1$$

이므로 기묘도도 보존된다. 중입자의 경우도 살펴보자. 사카타는 중입자가 다음과 같이 세 개의 기본 입자로 이루어진다고 생각했다.

$$\Sigma^+ = \Lambda p \overline{n}$$

$$\Xi^0 = \Lambda \Lambda \overline{n}$$

이 경우에도 아이소스핀의 z성분과 기묘도가 보존된다.

물리군 Λ의 기묘도가 -1이니까 기묘도가 -1인 강입자는 Λ가 한 개 들어가고 기묘도가 -2인 강입자는 Λ가 두 개 들어가는군요.

정교수 맞아.

쿼크모형의 창시자 겔만 _ 겔만의 생애와 연구

정교수   이제 쿼크 이론의 창시자인 겔만을 알아볼게.

겔만(Murray Gell-Mann, 1929~2019, 1969년 노벨 물리학상 수상, 사진 출처: World Economic Forum/Wikimedia Commons)

겔만은 미국 맨해튼에서 태어나 오스트리아-헝가리 제국 출신의 유대인 이민자 가정에서 자랐다. 소년 시절부터 그는 자연과 수학에 호기심이 가득했다. 그는 14세에 컬럼비아 문법 및 예비학교를 수석으로 졸업한 후 예일 대학에 입학했다. 예일대 재학 중에는 윌리엄 로웰 퍼트넘 수학 경시대회에 참가해 2등상을 수상했다.

1948년 겔만은 예일대에서 물리학 학사 학위를 받고, 대학원 공부를 위해 프린스턴 대학과 하버드 대학에 지원했다. 그는 프린스턴 대학에 떨어지고 하버드 대학에 합격했다. 하지만 하버드 대학은 재정 지원이 충분하지 않았다.

1930년대의 예일 대학

  결국 겔만은 매사추세츠 공과대학(MIT)에 입학 허가를 받았고, 이곳에서 조교를 하면서 필요한 재정을 지원받을 수 있었다. 그는 바이스코프(Victor Weisskopf) 교수의 지도하에 〈결합 강도와 핵반응〉이라는 제목의 논문을 완성하고, 1951년 MIT에서 물리학 박사 학위를 받았다.

  그 후 겔만은 프린스턴 고등연구소에서 박사 후 연구원으로 재직했다. 1952년부터 1953년까지는 일리노이 대학 어배너-섐페인 캠퍼스의 객원 연구 교수로 일했다. 1954년부터 1955년까지 그는 컬럼비아 대학의 객원 부교수와 시카고 대학의 부교수를 역임했다. 그리고 캘리포니아 공과대학으로 옮겨 1955년부터 1993년에 은퇴할 때까지 가르쳤다.

강입자의 정리 _ 짝지어 배열하기

정교수 물리학자들은 새롭게 발견된 중간자와 중입자를 여덟 개씩 짝을 지어 다음과 같은 방식으로 정리했어. 먼저 중간자의 경우 세 종류의 케이온과 중성 케이온의 반입자, 세 종류의 파이온과 에타 입자를 다음 그림과 같이 정육각형의 꼭짓점과 중심에 배치했지.

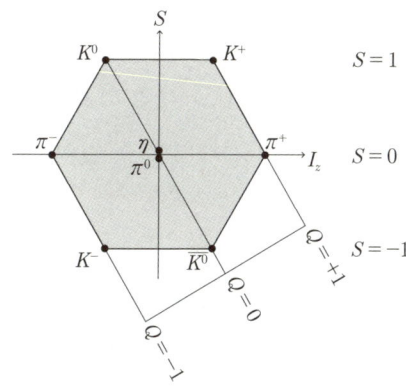

이렇게 배치하면 수평 방향의 입자끼리는 기묘도가 같고 수직 방향의 입자끼리는 I_z가 같아. 그리고 대각선 방향의 입자끼리는 전하가 같지.

물리군 중입자의 경우는 어떻게 8개의 입자를 배치하죠?

정교수 두 가지 방법이 있어. 첫 번째는 양성자, 중성자, 세 종류의 시그마 입자, 람다 입자, 두 종류의 크시 입자가 다음과 같이 배치되지.

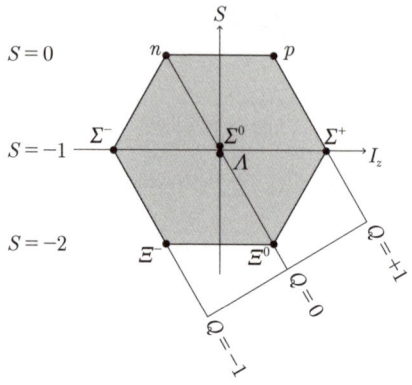

두 번째로는 네 개의 델타 입자와 세 개의 시그마 입자, 두 개의 크시 입자, 또 다른 입자를 볼링핀처럼 배열할 수 있어.

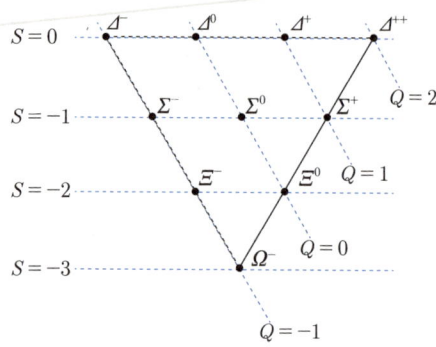

물리군 맨 아랫줄에 있는 Ω^-는 뭐예요?

정교수 당시에는 발견되지 않았고 나중에 겔만이 쿼크모형을 만들면서 예언한 입자야. 오메가 입자라고 부르지.

겔만의 쿼크모형 _제임스 조이스의 소설에서

정교수 1964년 겔만과 츠바이크는 독립적으로 양성자나 중성자 등의 강입자들이 세 개의 기본 입자로 이루어져 있다고 생각했어. 겔만은 이 기본 입자를 쿼크라고 불렀고, 츠바이크는 이 세 입자를 에이스라고 불렀어. 하지만 과학자들은 쿼크라는 이름을 더 좋아했지.

물리군 쿼크라는 이름은 어디에서 나온 건가요?

정교수 특별한 뜻은 없어. 겔만이 제임스 조이스의 소설《피네간의 경야》를 읽다가 거기서 나온 문장 속에서 찾은 단어야.

> 마크 대왕을 위한 세 개의 쿼크!
> 확실히 그는 대단한 규성(叫聲)은 갖지 않았나니
> 그리고 확실히 가진 것이라고는 모두 과녁(마크)을 빗나갔나니.
>
> — 제임스 조이스의 《피네간의 경야》

제임스 조이스(James Augustine Aloysius Joyce, 1882~1941)

겔만은 세 종류의 쿼크를 u쿼크, d쿼크, s쿼크라고 불렀어. 그는 s쿼크는 기묘도가 0이 아닌 강입자를 만드는 데 사용된다고 생각했어. s는 strange의 첫 글자야. 즉, 기묘도가 0이 아닌 기묘한 입자는 s쿼크를 가진다고 보았지. 그러니까 양성자와 중성자는 u쿼크와 d쿼크로만 구성돼.

$p = uud$

$n = udd$

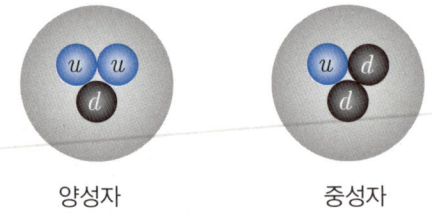

양성자　　　　　중성자

물리군　u쿼크와 d쿼크의 전하량은 어떻게 되나요?

정교수　u쿼크, d쿼크의 전하량을 각각 $Q(u)$, $Q(d)$라고 하면

$2Q(u) + Q(d) = 1$

$Q(u) + 2Q(d) = 0$

이야. 이 연립방정식을 풀면

$$Q(u) = +\frac{2}{3}$$

$$Q(d) = -\frac{1}{3}$$

이지.

물리군 전하량이 분수가 되는군요.

정교수 맞아.

물리군 두 쿼크의 아이소스핀의 z성분은요?

정교수 u쿼크, d쿼크의 아이소스핀의 z성분을 각각 $I_z(u)$, $I_z(d)$라고 하면 양성자의 아이소스핀의 z성분은 $\frac{1}{2}$이니까

$$\frac{1}{2} = 2I_z(u) + I_z(d)$$

이고, 중성자의 아이소스핀의 z성분은 $-\frac{1}{2}$이니까

$$-\frac{1}{2} = I_z(u) + 2I_z(d)$$

가 돼. 이 두 식을 연립하면

$$I_z(u) = \frac{1}{2}$$

$$I_z(d) = -\frac{1}{2}$$

이지. u쿼크는 업 상태를, d쿼크는 다운 상태를 나타내. 그래서 u쿼크

를 업 쿼크, d쿼크를 다운 쿼크라고 불러.

이번에는 델타 입자를 세 종류의 쿼크로 만들어 볼게. 델타 입자의 아이소스핀의 z성분과 전하량을 함께 쓰면 다음과 같아.

$\Delta^{++}: I_z = +\frac{3}{2} \quad Q = 2$

$\Delta^{+}: I_z = +\frac{1}{2} \quad Q = 1$

$\Delta^{0}: I_z = -\frac{1}{2} \quad Q = 0$

$\Delta^{-}: I_z = -\frac{3}{2} \quad Q = -1$

네 종류의 델타 입자는 기묘도가 0이므로 s쿼크는 안 들어가. 그러니까 u쿼크와 d쿼크로 다음과 같이 나타낼 수 있지.

$\Delta^{++} = uuu$

$\Delta^{+} = uud$

$\Delta^{0} = udd$

$\Delta^{-} = ddd$

예를 들면

$$I_z(\Delta^{++}) = 3I_z(u) = 3 \times \frac{1}{2} = \frac{3}{2}$$

$$Q(\Delta^{++}) = 3Q(u) = 3 \times \frac{2}{3} = 2$$

가 돼.

물리군 나머지는 제가 확인해 볼게요.

$$I_z(\Delta^+) = 2I_z(u) + I_z(d) = 2 \times \frac{1}{2} + \left(-\frac{1}{2}\right) = \frac{1}{2}$$

$$I_z(\Delta^0) = I_z(u) + 2I_z(d) = \frac{1}{2} + 2 \times \left(-\frac{1}{2}\right) = -\frac{1}{2}$$

$$I_z(\Delta^-) = 3I_z(d) = 3 \times \left(-\frac{1}{2}\right) = -\frac{3}{2}$$

이고,

$$Q(\Delta^+) = 2Q(u) + Q(d) = 2 \times \frac{2}{3} + \left(-\frac{1}{3}\right) = 1$$

$$Q(\Delta^0) = Q(u) + 2Q(d) = \frac{2}{3} + 2 \times \left(-\frac{1}{3}\right) = 0$$

$$Q(\Delta^-) = 3Q(d) = 3 \times \left(-\frac{1}{3}\right) = -1$$

이네요.

정교수 완벽해!

물리군 s쿼크의 전하량은 어떻게 되나요?

정교수 겔만은 s쿼크의 기묘도를 −1로 정했어. 그러니까 기묘도가 −1인 중입자에는 s쿼크가 한 개 들어가고, 기묘도가 −2인 중입자에는 s쿼크가 두 개 들어가지. 우선 기묘도가 −1인 중입자인 시그마 입자를 볼까? 이 입자의 전하량은 다음과 같아.

$$\Sigma^+ : Q = 1$$

$$\Sigma^0 : Q = 0$$

$$\Sigma^- : Q = -1$$

시그마 입자는 기묘도가 −1이니까 s쿼크가 한 개 들어가지. 나머지 두 개의 쿼크는 uu 또는 ud 또는 dd가 돼. 이 경우

$$(uu\text{의 전하량}) = \frac{2}{3} + \frac{2}{3} = \frac{4}{3}$$

$$(ud\text{의 전하량}) = \frac{2}{3} + \left(-\frac{1}{3}\right) = \frac{1}{3}$$

$$(dd\text{의 전하량}) = \left(-\frac{1}{3}\right) + \left(-\frac{1}{3}\right) = -\frac{2}{3}$$

야. 즉, s쿼크의 전하량은

$$-\frac{1}{3}$$

이고, 세 종류의 시그마 입자는 다음과 같이 만들 수 있어.

$\Sigma^+ = uus$

$\Sigma^0 = uds$

$\Sigma^- = dds$

마찬가지로 람다 입자와 크시 입자는 다음과 같이 구성되지.

$\Lambda = uds$

$\Xi^- = dss$

$\Xi^0 = uss$

오메가 입자는 기묘도가 −3이고 전하가 −1이어야 하니까 이 입자는 세 개의 s쿼크로 이루어져야 해.

$\Omega^- = sss$

물리군 중입자의 반입자는 어떤 쿼크로 구성되나요?

정교수 쿼크에 대한 반쿼크들로 구성되지. u쿼크, d쿼크, s쿼크의 반입자를 각각 $\bar{u}, \bar{d}, \bar{s}$로 나타내. 그러니까 반양성자는

$\bar{p} = \bar{u}\,\bar{u}\,\bar{d}$

이고, 반중성자는

$$\overline{n} = \overline{u}\,\overline{d}\,\overline{d}$$

가 돼.

물리군 그렇다면

$$\overline{\Sigma^-} = \overline{d}\,\overline{d}\,\overline{s}$$

이겠군요.

정교수 맞아. 중입자들은 세 개의 쿼크로 이루어져 있고, 중입자의 반입자는 세 개의 반쿼크로 이루어져 있지. 즉,

(중입자) $= qqq$

(중입자의 반입자) $= \overline{q}\,\overline{q}\,\overline{q}$

(여기서 $q = u, d, s$)

로 나타낼 수 있지.

물리군 겔만이 예언한 오메가 입자는 발견되었나요?

정교수 물론이야. 1964년 미국 브룩헤이븐 국립연구소(BNL)에서 AGS(Alternating Gradient Synchrotron)라는 입자가속기를 사용해 발견했어.

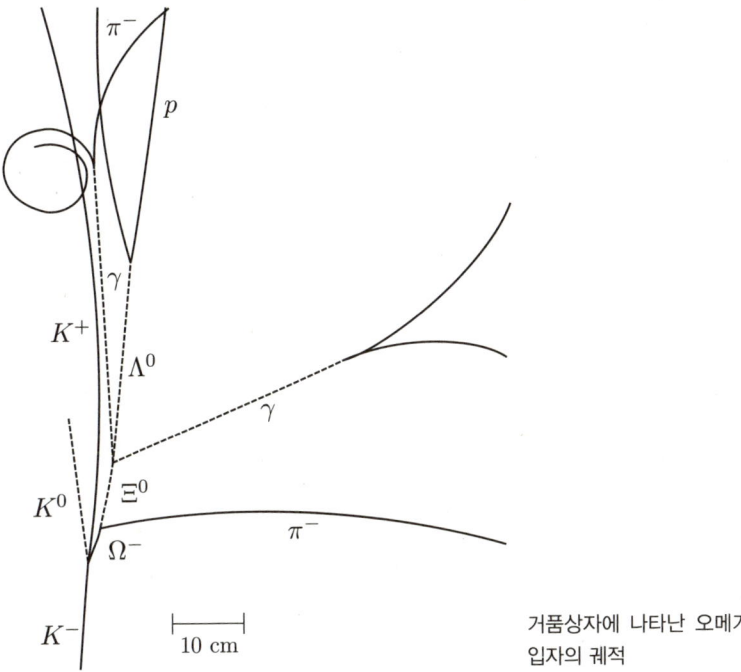

거품상자에 나타난 오메가
입자의 궤적

su(3)대수와 쿼크모형 _ 대칭군과 연결 짓다

정교수 앞에서 공부한 리군과 리대수를 이용해 쿼크모형을 설명해 볼게. 겔만은 다음과 같은 물음을 던졌어.

"양성자와 중성자는 매우 비슷한데 왜 질량이 조금 다를까?"

"$\Sigma^+, \Sigma^0, \Sigma^-$는 같은 계열처럼 보이지 않나?"

"어떤 대칭성이 숨어 있는 건 아닐까?"

그는 입자들이 수학적으로 어떤 대칭군과 관련있다고 생각했는데,

그게 바로 $SU(3)$군이었어. $SU(3)$군의 원소는 3×3 복소수 유니터리 행렬이고 행렬식이 1이지. 즉,

$U \in SU(3)$

이면

$UU^\dagger = I, \quad |U| = 1$

이 돼.

물리군 그건 왜죠?

정교수 $SU(2)$로 아이소스핀(양성자-중성자) 대칭을 설명했지만, s 쿼크를 포함하기엔 부족했어. 따라서 u, d, s 세 쿼크를 기저로 하는 3차원 복소 벡터 공간을 생각해야 했는데, 그게 바로 $SU(3)$군이었지.

여기서 $SU(3)$군에 대응하는 리대수인 $su(3)$대수를 생각하자. 겔만은 $su(3)$대수에 대한 8개의 생성자를 다음과 같이 행렬로 나타냈다.

$$\lambda_1 = \begin{pmatrix} 0 & 1 & 0 \\ 1 & 0 & 0 \\ 0 & 0 & 0 \end{pmatrix} \quad \lambda_2 = \begin{pmatrix} 0 & -i & 0 \\ i & 0 & 0 \\ 0 & 0 & 0 \end{pmatrix} \quad \lambda_3 = \begin{pmatrix} 1 & 0 & 0 \\ 0 & -1 & 0 \\ 0 & 0 & 0 \end{pmatrix} \quad \lambda_4 = \begin{pmatrix} 0 & 0 & 1 \\ 0 & 0 & 0 \\ 1 & 0 & 0 \end{pmatrix}$$

$$\lambda_5 = \begin{pmatrix} 0 & 0 & -i \\ 0 & 0 & 0 \\ i & 0 & 0 \end{pmatrix} \quad \lambda_6 = \begin{pmatrix} 0 & 0 & 0 \\ 0 & 0 & 1 \\ 0 & 1 & 0 \end{pmatrix} \quad \lambda_7 = \begin{pmatrix} 0 & 0 & 0 \\ 0 & 0 & -i \\ 0 & i & 0 \end{pmatrix} \quad \lambda_8 = \frac{1}{\sqrt{3}} \begin{pmatrix} 1 & 0 & 0 \\ 0 & 1 & 0 \\ 0 & 0 & -2 \end{pmatrix}$$

이 여덟 개의 행렬을 겔만행렬이라고 부른다.

겔만이 생각한 세 쿼크 상태는 다음과 같이 행렬로 나타낼 수 있다.

u쿼크 상태: $|u\rangle = \begin{pmatrix} 1 \\ 0 \\ 0 \end{pmatrix}$

d쿼크 상태: $|d\rangle = \begin{pmatrix} 0 \\ 1 \\ 0 \end{pmatrix}$

s쿼크 상태: $|s\rangle = \begin{pmatrix} 0 \\ 0 \\ 1 \end{pmatrix}$

겔만행렬에서 두 개의 대각행렬은 아이소스핀의 z성분 및 초전하와 관계된다.

$$I_z = \frac{1}{2}\lambda_3$$

$$Y = \frac{1}{\sqrt{3}}\lambda_8$$

그러므로

$$I_z|u\rangle = \frac{1}{2}|u\rangle$$

$$I_z|d\rangle = -\frac{1}{2}|d\rangle$$

$$I_z|s\rangle = 0|s\rangle$$

와

$$Y|u\rangle = \frac{1}{3}|u\rangle$$

$$Y|d\rangle = \frac{1}{3}|d\rangle$$

$$Y|s\rangle = -\frac{2}{3}|s\rangle$$

가 성립한다.

이제 다음과 같은 행렬을 생각하자.

$$T_+ = \frac{1}{2}(\lambda_1 + i\lambda_2) = \begin{pmatrix} 0 & 1 & 0 \\ 0 & 0 & 0 \\ 0 & 0 & 0 \end{pmatrix}$$

$$T_- = \frac{1}{2}(\lambda_1 - i\lambda_2) = \begin{pmatrix} 0 & 0 & 0 \\ 1 & 0 & 0 \\ 0 & 0 & 0 \end{pmatrix}$$

이 두 행렬은 u쿼크 상태와 d쿼크 상태의 변환을 나타낸다. 즉,

$$T_+|d\rangle = |u\rangle$$

$$T_-|u\rangle = |d\rangle$$

이다. 이제 다음과 같은 행렬을 생각하자.

$$V_+ = \frac{1}{2}(\lambda_4 + i\lambda_5) = \begin{pmatrix} 0 & 0 & 1 \\ 0 & 0 & 0 \\ 0 & 0 & 0 \end{pmatrix}$$

$$V_- = \frac{1}{2}(\lambda_4 - i\lambda_5) = \begin{pmatrix} 0 & 0 & 0 \\ 0 & 0 & 0 \\ 1 & 0 & 0 \end{pmatrix}$$

이 두 행렬은 u쿼크 상태와 s쿼크 상태의 변환을 나타낸다. 즉,

$$V_+ | s \rangle = | u \rangle$$

$$V_- | u \rangle = | s \rangle$$

이다. 이제 다음과 같은 행렬을 생각하자.

$$U_+ = \frac{1}{2}(\lambda_6 + i\lambda_7) = \begin{pmatrix} 0 & 0 & 0 \\ 0 & 0 & 1 \\ 0 & 0 & 0 \end{pmatrix}$$

$$U_- = \frac{1}{2}(\lambda_6 - i\lambda_7) = \begin{pmatrix} 0 & 0 & 0 \\ 0 & 0 & 0 \\ 0 & 1 & 0 \end{pmatrix}$$

이 두 행렬은 d쿼크 상태와 s쿼크 상태의 변환을 나타낸다. 즉,

$$U_+ | s \rangle = | d \rangle$$

$$U_- | d \rangle = | s \rangle$$

이다. 이 내용을 다이어그램으로 나타내면 다음 그림과 같다.

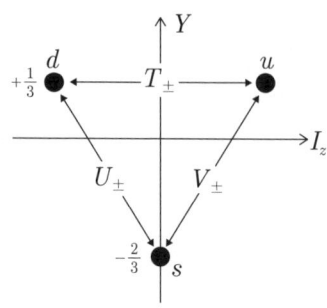

이렇게 겔만은 $su(3)$대수와 세 종류의 쿼크 사이의 관계를 완벽하게 묘사했다.

여섯 번째 만남

쿼크모형의 진화

한무영과 난부의 색깔이 있는 쿼크 _ 빛의 삼원색처럼

정교수 쿼크모형이 나온 후에 몇 가지 문제점이 제기되었어. 그중 하나는 파울리의 배타원리 문제야.

물리군 어떤 문제죠?

정교수 양성자는 세 개의 쿼크로 이루어져 있어. u쿼크 한 개와 d쿼크 두 개지. 파울리의 배타원리에서 업스핀을 가진 전자와 다운스핀을 가진 전자는 하나의 상태에 있을 수 있지만, 업스핀을 가진 전자 두 개가 하나의 상태에 있을 수는 없다고 얘기해. 물리학자들은 쿼크가 페르미온이라고 생각했어. 그러니까 u쿼크와 d쿼크는 하나의 상태에 있을 수 있지만, 두 개의 u쿼크 또는 두 개의 d쿼크는 하나의 상태에 있을 수 없잖아? 그렇다면 양성자

$$p = uud$$

는 허용되지 않아.

물리군 그럼 쿼크는 페르미온이 아닌가요?

정교수 그렇게 생각한 물리학자가 있어. 미국 메릴랜드 대학의 그린버그(Oscar Wallace Greenberg, 1932~)는 쿼크가 하나의 상태에 같은 종류가 세 개까지 허용되는 새로운 입자라고 주장했지. 그는 '여러 개'라는 뜻의 라틴어 파라(para)를 붙여서 이러한 법칙을 따르는 입자를 파라페르미온이라고 불렀어. 하지만 사람들은 그린버그의 생각에 큰 관심을 보이지 않았지. 이 문제를 다른 방법으로 해결한 사람이

대한민국의 한무영 교수과 일본의 난부야. 먼저 한무영에 대해 알아볼게.

한무영(1934~2016)은 1934년 대한민국 서울에서 태어났다. 그는 한국전쟁 이후 서울대학교 전기공학과를 1학년까지 다니고 미국 위스콘신주의 캐럴 대학에서 전기공학으로 학사 학위를 받았다. 그 후 로체스터 대학 물리학과 대학원에 진학해 1964년에 입자이론물리학으로 박사 학위를 받았다. 1967년에는 노스캐롤라이나주 더럼에 있는 듀크 대학의 물리학과 교수가 되었다.

이번에는 일본의 난부에 대해 알아보자.

난부 요이치로(南部陽一郎, 1921~2015, 2008년 노벨 물리학상 수상)

난부 요이치로는 1921년 일본 도쿄에서 태어났다. 그는 후쿠이 중학교를 졸업하고 도쿄 대학에서 물리학을 공부해 1942년에 학사 학위를 받았다. 이후 곧바로 육군에 입대해 레이더 관련 연구를 했다.

전쟁이 끝나자 난부는 다시 도쿄 대학으로 돌아왔다. 당시 도쿄 대학에는 양자장론을 강의할 교수가 없어서 수학과 교수인 고다이라 구니히코(小平邦彦, 1915~1997)[2]에게 강의를 부탁했다. 고다이라 교수는 학생 몇 명을 모아 양자장론 수업을 진행했다.

1948년 난부는 자신의 첫 논문을 발표했고, 이듬해에 두 번째 논문을 발표했다. 그는 아직 박사 과정 중임에도 불구하고 신설 대학인 오사카 시립대학의 부교수가 되었다. 이듬해에는 29세의 나이로 정교수가 되었다.

1952년 난부는 도쿄 대학에서 이학 박사 학위를 받았다. 그는 도모나가 신이치로의 추천으로 미국 프린스턴 고등연구소로 가게 되었다. 1954년부터는 시카고 대학의 연구원으로 지내다 1956년에 시카고 대학 물리학과 교수가 되었다.

물리군 한무영과 난부는 어떻게 문제를 해결했나요?

정교수 당시에는 논문을 완성하면 저널에 투고하기에 앞서, 타이핑한 원고인 프리프린트를 세계 각국의 주요 대학이나 연구소로 보내 자신이 한 연구를 알렸어. 1965년 한무영은 같은 쿼크 여러 개가 하나의 상태에 들어가기 위해서는 하나의 쿼크가 세 종류여야 한다고 주장했지. 그는 이 내용을 프리프린트로 만들어 각 대학에 보냈는데,

---

2) 1954년 일본 최초이자 아시아인 최초로 필즈상을 받은 수학자

시카고 대학의 난부도 마침 비슷한 생각을 하고 있었다네. 그들은 서로 만난 적은 없었지만 이 주제로 공동 연구를 하기로 했지. 그렇게 해서 두 사람이 쓴 논문이 세상에 알려졌어. 이것을 한-난부 모형이라고 불러.

훗날 겔만은 한-난부 모형의 세 종류의 쿼크를 빨강(R), 초록(G), 파랑(B)이라고 이름 붙였어.

물리군 왜 이 세 가지 색을 도입한 거죠?

정교수 빛의 삼원색이거든. 이 세 가지 색의 빛을 합성하면 흰색이 돼.

u쿼크에는 세 종류가 있다. 이것을 각각

$$u_R \quad u_G \quad u_B$$

로 쓴다. 마찬가지로 d쿼크와 s쿼크에도 다음과 같이 각각 세 종류의 쿼크가 있다.

$$d_R \quad d_G \quad d_B$$

$$s_R \quad s_G \quad s_B$$

한무영과 난부는 중입자가 세 개의 서로 다른 색 쿼크로 이루어져 있다고 생각했다. 예를 들어 양성자는 다음과 같다.

$$(양성자) = u_R \, u_B \, d_G$$

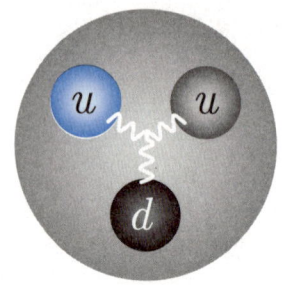

양성자의 쿼크 구조

두 개의 서로 다른 색의 u쿼크이므로 하나의 양성자 상태에 u쿼크 두 개가 들어갈 수 있다. 이것이 한-난부의 해결법이다. 또 다른 예로 중성자를 살펴보자.

$$(중성자) = u_B \, d_R \, d_G$$

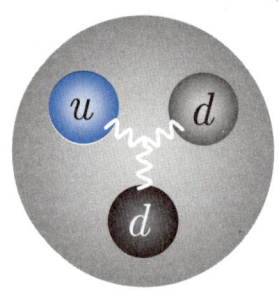

중성자의 쿼크 구조

마찬가지로

$$\Delta^{++} = u_R \, u_G \, u_B$$

$$\Delta^- = d_R \, d_G \, d_B$$

$$\Omega^- = s_R \, s_G \, s_B$$

이다.

물리군 쿼크의 반입자인 반쿼크도 세 가지 색이 있나요?

정교수 물론이야. 반쿼크는 $\bar{q}$ 라고 표시하는데 반빨강($\bar{R}$), 반초록($\bar{G}$), 반파랑($\bar{B}$)의 삼원색을 가지지. 그러니까 반쿼크는 다음과 같이 아홉 종류야.

$$\bar{u}_{\bar{R}} \quad \bar{u}_{\bar{G}} \quad \bar{u}_{\bar{B}}$$

$$\bar{d}_{\bar{R}} \quad \bar{d}_{\bar{G}} \quad \bar{d}_{\bar{B}}$$

$$\bar{s}_{\bar{R}} \quad \bar{s}_{\bar{G}} \quad \bar{s}_{\bar{B}}$$

예를 들어 반양성자와 반중성자는 다음과 같이 세 개의 반쿼크로 이루어져 있어.

$$(반양성자) = \bar{u}_{\bar{R}} \, \bar{u}_{\bar{B}} \, \bar{d}_{\bar{G}}$$

$$(반중성자) = \bar{u}_{\bar{B}} \, \bar{d}_{\bar{R}} \, \bar{d}_{\bar{G}}$$

이런 식으로 중입자는 세 개의 색이 다른 쿼크로 이루어져 있고, 중입자의 반입자는 세 개의 색이 다른 반쿼크로 이루어져 있지.

물리군   그럼 중입자는 세 가지 색이 섞여 색이 없는 상태가 되겠네요.

정교수   맞아. 흰색이 돼.

물리군   중간자는 어떤 색의 쿼크로 이루어져 있나요?

정교수   중간자도 중입자처럼 흰색이어야 해. 빨강과 반빨강을 섞으면 흰색이고, 마찬가지로 초록과 반초록 또는 파랑과 반파랑을 섞으면 흰색이야. 다시 말해 중간자 $q\bar{q}$ 는 $q_R \bar{q}_R$ 와 $q_G \bar{q}_G$ 와 $q_B \bar{q}_B$ 로 이루어져 있어. 이렇게 섞으면 중간자는 흰색이지.

물리군   강입자는 모두 흰색이군요.

정교수   그리고 한 가지 더. 한무영과 난부는 겔만이 주장한 분수 전하를 싫어했어. 두 사람은 쿼크 역시 정수 전하를 가져야 한다고 생각했지. 쿼크의 전하량을 Q라고 하면 한-난부 모형에서의 전하량은 다음과 같아.

$Q(u_R) = 1$

$Q(u_G) = 1$

$Q(u_B) = 0$

$Q(d_R) = 0$

$Q(d_G) = 0$

$Q(d_B) = -1$

$Q(s_R) = 0$

$Q(s_G) = 0$

$Q(s_B) = -1$

한-난부는 각 색깔 쿼크 전하의 평균이 겔만이 얘기한 분수 전하가 된다고 보았어.

(겔만의 u쿼크 전하) $= \frac{1}{3}(Q(u_R) + Q(u_G) + Q(u_B)) = \frac{2}{3}$

(겔만의 d쿼크 전하) $= \frac{1}{3}(Q(d_R) + Q(d_G) + Q(d_B)) = -\frac{1}{3}$

(겔만의 s쿼크 전하) $= \frac{1}{3}(Q(s_R) + Q(s_G) + Q(s_B)) = -\frac{1}{3}$

물리군 겔만도 쿼크의 색깔을 인정했나요?

정교수 물론이야. 하지만 겔만은 세 가지 색깔 쿼크의 전하량을 분수로 선택했어. 그러니까 다음과 같아.

$Q(u_R) = Q(u_G) = Q(u_B) = \frac{2}{3}$

$Q(d_R) = Q(d_G) = Q(d_B) = -\frac{1}{3}$

$Q(s_R) = Q(s_G) = Q(s_D) = -\frac{1}{3}$

과학자들은 한-난부의 정수 전하보다는 겔만의 분수 전하를 더 올바른 쿼크모형으로 인정했다네.

참 쿼크의 존재를 예언한 과학자들 _ 비요르켄, 일리오풀로스, 마이아니, 이휘소

정교수 이번에는 네 번째 쿼크인 참 쿼크를 예언한 과학자들을 살펴볼 거야. 먼저 비요르켄에 대해 알아볼게.

비요르켄(James Daniel "BJ" Bjorken, 1934~2024)

비요르켄은 스웨덴 출신 이민자 집안에서 태어났다. 그의 아버지는 스웨덴을 떠나 미국 시카고에서 전기 엔지니어로 일했다. 비요르켄은 시카고에서 자랐으며 어릴 때부터 수학과 과학을 좋아했다. 그는 1952년 메인이스트 고등학교를 졸업한 후 시카고 대학을 거쳐 매사추세츠 공과대학(MIT)에서 물리학 학사를, 1959년 스탠퍼드 대학에서 박사 학위를 받았다. 그는 스탠퍼드 선형가속기 센터(SLAC)의 이론 그룹 교수였으며, 페르미 국립 가속기 연구소의 이론 부서 구성원이었다.

두 번째 소개할 과학자는 일리오풀로스다.

일리오풀로스(John Iliopoulos, 1940~, 사진 출처: Binant/Wikimedia Commons)

일리오풀로스는 1962년 그리스 아테네 국립기술대학(NTUA) 기계 전기 공학과를 졸업했다. 그는 파리 대학에서 이론물리학 연구를 계속했으며, 1968년에 박사 학위를 받았다. 1969년부터 1971년까지는 하버드 대학 연구원으로 재직했다. 또한 1991~1995년과 1998~2002년에 파리 고등사범학교 이론물리학 연구소의 소장직을 역임했다.

세 번째로 소개할 물리학자는 마이아니다.

마이아니(Luciano Maiani, 1941~,
사진 출처: Laurent Guiraud/Wikimedia Commons)

1964년 루치아노 마이아니는 물리학 학위를 받고 이탈리아 국립 보건원의 연구원이 되었다. 같은 해에 피렌체 대학 라울 가토(Raoul Gatto)의 이론물리학 그룹과 공동 연구를 했다. 1969년 그는 대서양을 건너 하버드의 라이먼 물리학 연구소에서 박사 후 연구원 과정을 밟았다.

1976년 마이아니는 로마 대학의 이론물리학 교수가 되었다. 또한 같은 기간에 파리 고등사범학교(1977)와 CERN(1979~1980 및 1985~1986)에서 객원 교수직을 맡았다.

네 번째로 소개할 과학자는 대한민국의 이휘소 교수이다.

이휘소는 대한민국 서울 용산에서 사 남매 중 맏이로 태어났다. 그의 부

이휘소(Benjamin Whisoh Lee, 1935~1977)

모는 모두 의학 교육을 받다. 그의 어머니는 집안의 생계를 책임지고 있었다. 그녀는 병원에서 의사로 일하다가 나중에 소아과 및 산부인과 진료소를 개업했다.

이휘소는 어렸을 때부터 학업에 재능을 보이며 경기중학교에 입학했다. 4년째 되던 해 한국전쟁이 발발해 가족들과 부산 외곽으로 피난을 가야 했다. 그곳에서 그는 학업을 계속했다.

이후 경기고등학교에 입학한 이휘소는 졸업을 1년 앞두고 서울대학교 화학공학과에 수석으로 입학했다. 대학 재학 중 한국전쟁 참전 군인 부인회에서 장학금을 받아 미국으로 이민을 가서 학부 과정을 밟을 수 있었다.

이휘소는 1956년 마이애미 대학에서 최우등으로 학사 학위를, 1958년 피츠버그 대학에서 석사 학위를, 1961년 펜실베이니아 대학에서 박사 학위를 받았다. 그는 프린스턴 고등연구소에서 연구를 수행한 뒤 펜실베이니아 대학, 스토니브룩 대학, 시카고 대학에서 물리학 교수로 재직했다. 이후 페르미 국립 가속기 연구소의 이론물리학과 학과장으로 임명되었다.

1964년 이휘소는 지도교수 에이브러햄 클라인(Abraham Klein)과 함께 자발적 대칭 파괴에 관한 논문을 발표했는데, 이것이 힉스 메커니즘의 출현에 기여했다. 1969년에 그는 자발적으로 깨진 게이지 대칭을 재정규화하는 데 성공했다. 또한 K중간자의 혼합과 붕괴에 해당하는 양을 계산해 참 쿼크의 질량을 예측했다.

1977년 6월 16일, 이휘소는 일리노이주 키와니 인근에서 80번 주

간고속도로를 달리던 중 교통사고로 사망했다.

참 쿼크의 필요성 _ 입자가속기의 실험 결과로부터

정교수 이제 참 쿼크의 존재를 주장한 과학자들의 아이디어를 살펴볼게.

겔만이 세 종류의 쿼크를 도입했을 때 양의 전기를 띤 쿼크는 u쿼크 한 종류이고, 음의 전기를 띤 쿼크는 d쿼크와 s쿼크의 두 종류였다. 과학자들은 이들의 짝이 맞지 않는다고 생각했다. u쿼크와 같은 전하를 갖는 새로운 쿼크가 존재해서 그것이 s쿼크와 짝을 이뤄야 한다고 보았다. 그래서 s를 나타내는 strange의 반대 개념으로 charm이라는 단어를 도입했다. 그리고 이 쿼크의 이름을 참 쿼크(charm quark)라 하고 c로 나타내기로 했다. 참 쿼크의 전하량은 $+\frac{2}{3}$이고 다음과 같이 두 쌍의 쿼크 조합이 생긴다.

(u, d)

(c, s)

1964년 비요르켄과 글래쇼는 당시 네 종류의 경입자(전자, 뮤온,

그리고 각각의 뉴트리노)가 있으므로 쿼크도 네 종류이어야 한다고 생각했다. 1970년 글래쇼, 일리오풀로스, 마이아니는 네 번째 쿼크인 참 쿼크를 실험으로 찾을 방법을 제시했다.

당시 참 쿼크의 필요성을 알리는 실험 결과가 나왔다. 광자가 쿼크 쌍을 만드는 과정의 확률은 각 쿼크 전하량의 제곱의 합에 비례한다. 각 쿼크 전하량의 제곱의 합을 R이라고 하자. 만일 겔만이 얘기한 대로 쿼크가 세 종류라면 각 쿼크당 세 종류의 쿼크가 있으므로 쿼크 9개의 전하량의 제곱의 합이 R이다. 따라서

$$R = 3 \times \left[\left(\frac{2}{3}\right)^2 + \left(-\frac{1}{3}\right)^2 + \left(-\frac{1}{3}\right)^2\right] = 2$$

이다.

이탈리아의 입자가속기 ADONE에서 충돌 에너지가 1~3GeV일 때 $R = 2$가 되어 겔만의 쿼크모형에 잘 맞았다. 하지만 케임브리지 전자가속기에서 충돌 에너지를 4~5GeV로 올렸을 때 R의 값은 2보다 커졌다. 이것은 새로운 쿼크가 존재해야 함을 의미했다.

브룩헤이븐 연구소를 방문한 글래쇼는 연구소 책임자인 사미오스에게 뉴트리노와 양성자의 충돌 과정을 잘 분석해 보면 참 쿼크의 증거를 찾을 수 있을 거라고 설명했다.

W^+입자의 교환으로 양성자 속의 d쿼크가 u쿼크로 변하면서 아주 짧은 시간 동안 참 쿼크를 포함하는 숨입자가 생성될 거예요. 이것을

식으로 쓰면

$$d + W^+ \to u$$

가 되지요. 이 중입자는 순식간에 붕괴되는데 이 과정에서 람다 입자 Λ^0도 만들어집니다. 이때 람다 입자 속의 s쿼크는 W^-입자의 교환으로 참 쿼크가 변한 것일 수 있습니다. 즉,

$$c + W^- \to s$$

와 같이요. 그러니까 Λ^0이 나타나는 V자 모양의 궤적을 찾으면 참 쿼크가 존재할 가능성이 크지요.

- 글래쇼

글래쇼와 헤어진 사미오스는 연구소의 사진들을 뒤지기 시작했다. 그리고 1974년 5월 말에 다음과 같은 사진을 찾아냈다.

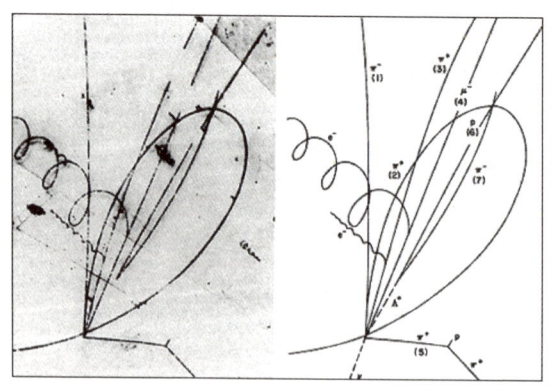

이 사진은 뉴트리노가 양성자와 충돌한 뒤 파이온을 비롯한 입자들이 생성되는 과정에서 Λ^0이 나타나는 V자 모양의 궤적을 보여준다. 이는 글래쇼가 찾던 사진이었다. 하지만 이것이 참 쿼크의 존재를 증명한다고 볼 수는 없었다.

글루온 _ 마치 고무줄처럼

정교수 핵을 구성하는 소립자가 강입자에서 쿼크로 바뀌면서 강력의 실체는 쿼크와 쿼크 사이의 상호작용이라는 것이 알려졌어.

물리군 쿼크와 쿼크 사이의 힘을 매개하는 입자는 뭐죠?

정교수 이 입자는 글루온이라고 불러. 아직까지 발견되지는 않았지. 글루온은 쿼크를 그룹으로 묶어 양성자 및 중성자와 같은 강입자를 만드는 역할을 해.

물리군 그럼 전자기력에서 광자처럼, 강력에서는 글루온이 힘의 매개 입자인가요?

정교수 정확해. 쿼크와 쿼크 사이의 강한 상호작용을 글루온이 전달하지. 이 강한 힘에는 아주 특별한 성질이 있어. 이름하여 점근적 자유(asymptotic freedom)라는 거야.

물리군 그게 뭐예요?

정교수 말 그대로 쿼크들이 아주 가까이 있을 땐 자유롭게 움직이지만, 멀어지려고 하면 점점 더 강한 힘이 작용해서 도망칠 수 없는 성질이야.

물리군 오! 마치 고무줄처럼 가까이 있을 땐 느슨하지만, 멀어지면 팽팽하게 당겨지는 느낌이네요.

정교수 적절한 비유야. 그 때문에 우리는 쿼크를 절대 단독으로 발견할 수 없어. 멀어지면 글루온의 결속력이 너무 강해져서 결국 새로운 강입자가 생기지.

물리군 그래서 '쿼크는 갇혀 있다'는 표현을 쓰는 거군요.

정교수 맞아. 그걸 색깔의 감금(color confinement)이라고도 불러.

물리군 근데 글루온은 아직도 발견이 안 됐다고 하셨죠?

정교수 그래. 자유로운 글루온은 아직 직접 검출되지 않았어. 하지만 입자 충돌 실험에서의 제트(jet) 현상이나, 강한 상호작용의 행동을 통해 그 존재는 거의 확실하다고 보고 있어.

물리군 글루온이 쿼크들을 묶고, 점근적 자유 때문에 서로 멀어질 수 없고……. 결국 우리가 알고 있는 양성자나 중성자 같은 입자가 만들어지는 거네요.

정교수 그렇지.

물리군 점근적 자유에 대한 이론은 누가 처음 생각했나요?

정교수 세 명의 물리학자가 생각했어. 바로 그로스, 윌첵, 폴리처라네.

그로스(David Jonathan Gross, 1941~, 2004년 노벨 물리학상 수상, 사진 출처: LANL)

윌첵(Frank Anthony Wilczek, 1951~, 2004년 노벨 물리학상 수상, 사진 출처: Kenneth C. Zirkel/Wikimedia Commons)

폴리처(Hugh David Politzer, 1949~, 2004년 노벨 물리학상 수상, 사진 출처: 미국물리학회(AIP))

참 쿼크의 발견 _ 독립적으로 찾아낸 새로운 중간자

정교수 이번에는 참 쿼크를 발견한 두 명의 과학자를 소개할게. 먼저 미국의 팅에 대해 알아볼까?

팅(Samuel Chao Chung Ting, 1936~,
1976년 노벨 물리학상 수상,
사진 출처: AlexHe34/Wikimedia Commons)

팅은 1936년 미국 미시간주 앤아버에서 산둥성 주현 출신의 중국계 이민 1세대 부모에게서 태어났다. 팅의 부모는 그가 태어난 지 두 달 후에 중국으로 돌아갔고, 팅은 제2차 세계대전 동안 부모로부터 홈스쿨링을 받았다. 공산당이 중국 본토를 장악한 후 국민당 정부는 타이완으로 도망쳐야 했기 때문에, 팅의 가족도 1949년에 타이완으로 이주했다. 팅은 1949년부터 1956년까지 타이완에서 살았으며, 대부분의 정규 학교 교육을 타이완에서 마쳤다. 그는 타이베이의 청쿵 고등학교를 졸업하고 국립 청쿵 대학에 입학해 한 학기 동안 공부했다.

1956년 영어를 거의 할 줄 몰랐던 팅은 20세의 나이에 미국으로 가서 미시간 대학에 다녔다. 그곳에서 공학, 수학, 물리학을 공부했다. 그는 1959년에 수학 및 물리학, 공학 학사, 1960년에 물리학 석사, 1962년에 물리학 박사 학위를 받았다.

1963년 팅은 유럽 입자물리 연구소(CERN)에서 근무했다. 그는 1965년부터 뉴욕의 컬럼비아 대학에서 강의했으며, 독일의 DESY(Deutsches Elektronen-Synchrotron)에서 근무했다. 1969년부터는 매사추세츠 공과대학(MIT)의 교수로 재직했다.

두 번째로 소개할 과학자는 미국의 릭터이다.

릭터(Burton Richter, 1931~2018, 1976년 노벨 물리학상 수상)

릭터는 뉴욕 브루클린의 유대인 가정에서 태어나 퀸스의 파로커웨이 지역에서 자랐다. 그는 파로커웨이 고등학교를 졸업했는데, 이곳은 리처드 파인먼(Richard Feynman)을 배출한 학교다. 릭터는 펜실베이니아의 머서스버그 아카데미에 다녔고, 그 후 매사추세츠 공과대학에서 공부해 1952년에 학사 학위를, 1956년에 박사 학위를 받았다. 그리고 스탠퍼드 대학 교수진에 합류해 1967년에 정교수가 되었다. 릭터는 1984년부터 1999년까지 스탠퍼드 선형가속기 센터(SLAC)의 소장을 역임하면서 SPEAR(Stanford Positron-Electron Asymmetric Ring) 입자가속기를 설계했다.

물리군 두 사람이 공동 연구를 했나요?
정교수 아니. 독립적으로 참 쿼크를 발견했어.

두 사람이 참 쿼크 발견에 본격적으로 뛰어든 것은 1974년이었다. 팅은 브룩헤이븐에 있는 양성자 가속기를 이용해 전자와 양전자를 검출해 새로운 중간자를 찾는 실험을 했다. 릭터는 스탠퍼드 선형가속기 센터에서 양전자-전자 충돌기를 이용해 새로운 중간자를 찾으려고 했다.

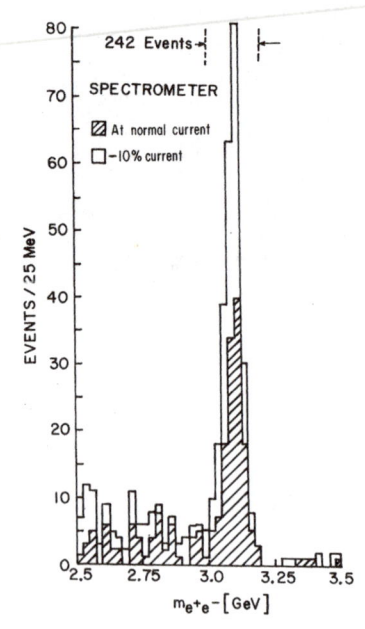

1974년 10월 13일, 팅은 새로운 중간자를 31억 전자볼트(3.1GeV)에서 발견했다. 팅은 이것을 J입자라고 불렀다.

릭터도 같은 해 10월에 새로운 중간자를 31억 전자볼트(3.1GeV)에서 발견했다. 릭터는 이것을

ψ(프시)입자로 불렀다.

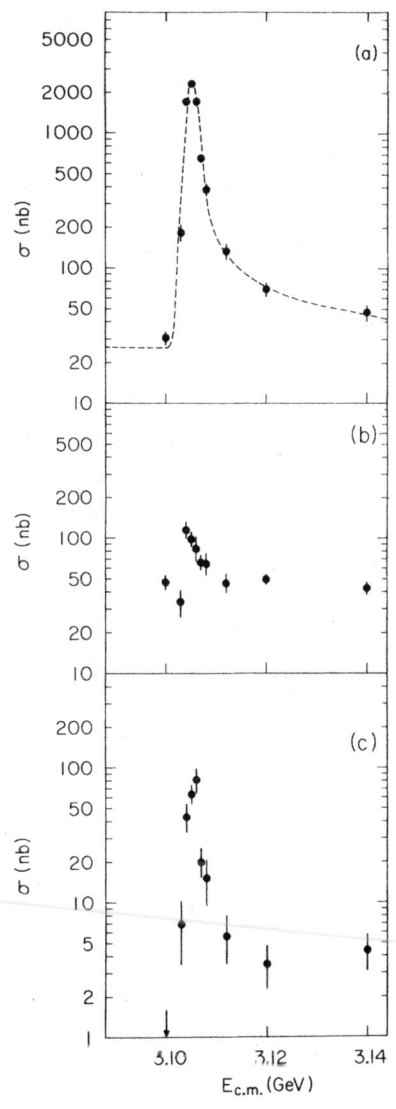

팅은 발견 내용을 11월 12일에 《Physical Review Letters》라는 저널에 투고했다. 릭터는 하루 뒤인 11월 13일에 자신의 결과를 같은 저널에 투고했다. 두 논문은 나란히 저널에 실렸다.

두 사람이 발견한 새로운 중간자는 참 쿼크(c)와 참 쿼크의 반쿼크($\bar{c}$)로 이루어진 중간자로 이름은 J/ψ입자로 명명되었다. 이 입자의 정지에너지는 3.1GeV이다. 이 발견으로 팅과 릭터는 1976년 노벨 물리학상을 공동 수상했다.

그 후 1976년 참 쿼크를 포함하는 새로운 중간자인 D중간자가 SLAC에서 발견되었다.

$$D^+ = c\bar{d} \quad (정지에너지) = 1.87\text{GeV}$$

$$D^0 = c\bar{u} \quad (정지에너지) = 1.86\text{GeV}$$

$$D_s^+ = c\bar{s} \quad (정지에너지) = 1.97\text{GeV}$$

보텀 쿼크와 톱 쿼크 _ 이론에서 증거 수집을 거쳐 발견에 이르기까지

물리군 발견된 쿼크는 네 종류인가요?

정교수 아니. 모두 여섯 종류야. 보텀 쿼크(bottom quark)와 톱 쿼크(top quark)가 있거든.

보텀 쿼크는 1973년 물리학자 고바야시 마코토(小林誠)와 마스카와 도시히데(益川敏英)가 처음 이론적으로 도입했다.

보텀 쿼크에 대한 증거는 1977년 리언 레더먼(Leon M. Lederman)이 이끄는 페르미랩(Fermilab) E288 실험 팀이 최초로 얻었다. 보텀 쿼크의 정지에너지는 약 4.18GeV로 양성자 정지에너지의 4배 이상이다.

톱 쿼크를 찾기 위한 SLAC(미국 스탠퍼드 대학)와 DESY(독일 함부르크)의 시도는 실패로 돌아갔다. 테바트론은 2009년 CERN의 LHC 운영 전까지 톱 쿼크를 생산할 만큼 강력하고 유일한 강입자 충돌기였다. 이미 존재하는 페르미랩의 Collider Detector(CDF) 외에 두 번째 검출기인 DØ 검출기가 복합체에 추가되었다. 1994년 4월 22일, CDF 그룹은 약 175GeV의 정지에너지를 가진 톱 쿼크의 존재에 대한 잠정적인 증거를 제시하는 논문을 제출했다. 한편 DØ가 더 많은 증거를 수집하고 DØ 데이터를 재분석한 후 1년이 지난 1995년 3월 2일, 두 그룹은 공동으로 176GeV의 정지에너지를 가진 톱 쿼크를 발견했다.

만남에 덧붙여

THE PRODUCTION OF HIGH SPEED LIGHT IONS WITHOUT THE USE OF HIGH VOLTAGES

BY ERNEST O. LAWRENCE AND M. STANLEY LIVINGSTON

UNIVERSITY OF CALIFORNIA

(Received February 20, 1932)

ABSTRACT

The study of the nucleus would be greatly facilitated by the development of sources of high speed ions, particularly protons and helium ions, having kinetic energies in excess of 1,000,000 volt-electrons; for it appears that such swiftly moving particles are best suited to the task of nuclear excitation. The straightforward method of accelerating ions through the requisite differences of potential presents great experimental difficulties associated with the high electric fields necessarily involved. The present paper reports the development of a method that avoids these difficulties by means of the multiple acceleration of ions to high speeds without the use of high voltages. The method is as follows: Semi-circular hollow plates, not unlike duants of an electrometer, are mounted with their diametral edges adjacent, in a vacuum and in a uniform magnetic field that is normal to the plane of the plates. High frequency oscillations are applied to the plate electrodes producing an oscillating electric field over the diametral region between them. As a result during one half cycle the electric field accelerates ions, formed in the diametral region, into the interior of one of the electrodes, where they are bent around on circular paths by the magnetic field and eventually emerge again into the region between the electrodes. The magnetic field is adjusted so that the time required for traversal of a semi-circular path within the electrodes equals a half period of the oscillations. In consequence, when the ions return to the region between the electrodes, the electric field will have reversed direction, and the ions thus receive second increments of velocity on passing into the other electrode. Because the path radii within the electrodes are proportional to the velocities of the ions, the time required for a traversal of a semi-circular path is independent of their velocities. Hence if the ions take exactly one half cycle on their first semi-circles, they do likewise on all succeeding ones and therefore spiral around in resonance with the oscillating field until they reach the periphery of the apparatus. Their final kinetic energies are as many times greater than that corresponding to the voltage applied to the electrodes as the number of times they have crossed from one electrode to the other. This method is primarily designed for the acceleration of light ions and in the present experiments particular attention has been given to the production of high speed protons because of their presumably unique utility for experimental investigations of the atomic nucleus. Using a magnet with pole faces 11 inches in diameter, a current of 10^{-9} ampere of 1,220,000 volt-protons has been produced in a tube to which the maximum applied voltage was only 4000 volts. There are two features of the developed experimental method which have contributed largely to its success. First there is the focussing action of the electric and magnetic fields which prevents serious loss of ions as they are accelerated. In consequence of this, the magnitudes of the high speed ion currents obtainable in this indirect manner are comparable with those conceivably obtainable by direct high voltage methods. Moreover, the focussing action results in the generation of very narrow beams of ions—less than 1 mm cross-sectional diameter—which are ideal for experimental studies of collision processes. Of hardly less importance is the second feature of the method which is the simple and highly effective means for the correction of the magnetic field along the paths of the ions. This makes it possible, indeed easy, to operate the tube effectively with a very high amplification factor (i.e., ratio of final equivalent voltage of accelerated ions to applied voltage). In consequence, this method in its present stage of development constitutes a highly reliable and experimentally convenient source of high speed ions requiring relatively modest laboratory equipment. Moreover, the present experiments indicate that this indirect method of multiple acceleration now makes practicable the production in the laboratory of protons having kinetic energies in excess of 10,000,000 volt-electrons. With this in mind, a magnet having pole faces 114 cm in diameter is being installed in our laboratory.

INTRODUCTION

THE classical experiments of Rutherford and his associates[1] and Pose[2] on artificial disintegration, and of Bothe and Becker[3] on excitation of nuclear radiation, substantiate the view that the nucleus is susceptible to the same general methods of investigation that have been so successful in revealing the extra-nuclear properties of the atom. Especially do the results of their work point to the great fruitfulness of studies of nuclear transitions excited artificially in the laboratory. The development of methods of nuclear excitation on an extensive scale is thus a problem of great interest; its solution is probably the key to a new world of phenomena, the world of the nucleus.

But it is as difficult as it is interesting, for the nucleus resists such experimental attacks with a formidable wall of high binding energies. Nuclear energy levels are widely separated and, in consequence, processes of nuclear excitation involve enormous amounts of energy—millions of volt-electrons.

It is therefore of interest to inquire as to the most promising modes of nuclear excitation. Two general methods present themselves; excitation by absorption of radiation (gamma radiation), and excitation by intimate nuclear collisions of high speed particles.

Of the first it may be said that recent experimental studies [4,5] of the absorption of gamma radiation in matter show, for the heavier elements, variations with atomic number that indicate a quite appreciable nuclear effect. This suggests that nuclear excitation by absorption of radiation is perhaps a not infrequent process, and therefore that the development of an intense artificial source of gamma radiation of various wave-lengths would be of considerable value for nuclear studies. In our laboratory, as elsewhere, this being attempted.

But the collision method appears to be even more promising, in consequence of the researches of Rutherford and others cited above. Their pioneer investigations must always be regarded as really great experimental achievements, for they established definite and important information about nuclear processes of great rarity excited by exceedingly weak beams of bombarding particles—alpha-particles from radioactive sources. Moreover, and this is the point to be emphasized here, their work has shown strikingly the

---

[1] See Chapter 10 of Radiations from Radioactive Substances by Rutherford, Chadwick and Ellis.
[2] H. Pose, Zeits. f. Physik 64, 1 (1930).
[3] W. Bothe and H. Becker, Zeits. f. Physik 66, 1289 (1930).
[4] G. Beck, Naturwiss. 18, 896 (1930).
[5] C. Y. Chao, Phys. Rev. 36, 1519 (1930).

great fruitfulness of the kinetic collision method and the importance of the development of intense artificial sources of alpha-particles. Of course it cannot be inferred from their experiments that alpha-particles are the most effective nuclear projectiles: the question naturally arises whether lighter or heavier particles of given kinetic energy would be more effective in bringing about nuclear transitions.

A beginning has been made on the theoretical study of the nucleus and a partial answer to this question has been obtained. Gurney and Condon[6] and Gamow[7] have independently applied the ideas of the wave mechanics to radioactivity with considerable success. Gamow[8] has further considered along the same lines the penetration into the nucleus of swiftly moving charged particles (with excitation of nuclear transitions in mind) and has concluded that, for a given kinetic energy, the lighter the particle the greater is the probability that it will penetrate the nuclear potential wall. This result is not unconnected with the smaller momentum and consequent longer wavelength of the ligher particles; for it is well-known that transmission of matter waves through potential barriers becomes greater with increasing wavelengths.

If the probability of nuclear excitation by a charged particle were mainly dependent on its ability to penetrate the nuclear potential wall, electrons would be the most effective. However, there is considerable evidence that nuclear excitation by electrons is negligible. It suffices to mention here the current view that the average density of the extra-nuclear electrons is quite great in the region of the nucleus, i.e., that the nucleus is quite transparent to electrons; in other words, there are no available stable energy levels for them.

On the other hand, there is evidence that there are definite nuclear levels for protons as well as alpha-particles;[9] indeed, there is some justification for the view that the general principles of the quantum mechanics are applicable in the nucleus to protons and alpha particles. It is not possible at the present time to estimate the relative excitation probabilities of the protons and alpha particles that succeed in penetrating the nucleus. However, it does seem likely that the greater penetrability of the proton* is an advantage outweighing any differences in their excitation characteristics. Protons thus appear to be most suited to the task of nuclear excitation.

Though at present the relative efficacy of protons and alpha-particles cannot be established with much certainty, it does seem safe to conclude at least that the most efficacious nuclear projectiles will prove to be swiftly moving ions, probably of low atomic number. In consequence it is important to develop methods of accelerating ions to speeds much greater than have heretofore been produced in the laboratory.

[6] Gurney and Condon, Phys. Rev. **33**, 127 (1929).
[7] Gamow, Zeits. f. Physik **51**, 204 (1928).
[8] Gamow, Zeits. f. Physik **52**, 514 (1929).
[9] J. Chadwick, J. E. R. Constable, E. C. Pollard, Proc. Roy. Soc. **A130**, 463 (1930).

* According to Gamow's theory a one million volt-proton has as great a penetrating power as a sixteen million volt alpha-particle.

The importance of this is generally recognized and several laboratories are developing techniques of the production and the application to vacuum tubes of high voltages for the generation of high speed electrons and ions. Highly significant progress in this direction has been made by Coolidge,[10] Lauritsen,[11] Tuve, Breit, Hafstad, Dahl,[12] Brasch and Lange,[13] Cockroft and Walton,[14] Van de Graaff[15] and others, who have developed several distinct techniques which have been applied to voltages of the order of magnitude of one million.

These methods involving the direct utilization of high voltages are subject to certain practical limitations. The experimental difficulties go up rapidly with increasing voltage; there are the difficulties of corona and insulation and also there is the problem of design of suitable high voltage vacuum tubes.

Because of these difficulties we have thought it desirable to develop methods for the acceleration of charged particles that do not require the use of high voltages. Our objective is two fold: first, to make the production of particles having kinetic energies of the order of magnitude of one million volt-electrons a matter that can be carried through with quite modest laboratory equipment and with an experimental convenience that, it is hoped, will lead to a widespread attack on this highly important domain of physical phenomena; and second, to make practicable the production of particles having kinetic energies in excess of those producible by direct high voltage methods—perhaps in the range of 10,000,000 volt-electrons and above.

A method for the multiple acceleration of ions to high speeds, primarily designed for heavy ions, has recently been described in this journal.[16] The present paper is a report of the development of a method for the multiple acceleration of light ions.[17] Particular attention has been given to the acceleration of protons because of their apparent unique utility in nuclear studies. In the present work relatively large currents of 1,220,000 volt-protons have been generated and there is foreshadowed in the not distant future the production of 10,000,000 volt-protons.

The Experimental Method

In the method for the multiple acceleration of ions to high speeds, recently described,[16] the ions travel through a series of metal tubes in synchronism with an applied oscillating electric potential. It is so arranged that as an

---

[10] W. D. Collidge, Am. Inst. E. Eng. **47**, 212 (1928).

[11] C. C. Lauritsen and R. D. Bennett, Phys. Rev. **32**, 850 (1928).

[12] M. A. Tuve, G. Breit, L. R. Hafstad and O. Dahl, Phys. Rev. **35**, 66 (1930); M. A. Tuve, L. R. Hafstad, O. Dahl, Phys. Rev. **39**, 384, (1932).

[13] A. Brasch and J. Lange, Zeits. f. Physik **70**, 10 (1931).

[14] J. J. Cockroft and E. T. S. Walton, Proc. Roy. Soc. A**129**, 477 (1930).

[15] R. S. Van de Graaff, Schenectady Meeting American Physical Society, 1931.

[16] D. H. Sloan and E. O. Lawrence, Phys. Rev. **38**, 2021 (1931).

[17] This method was first described before the September, 1930, meeting of the National Academy of Sciences (Lawrence and Edlefsen, Science **72**, 376-377 (1930)). Later before the American Physical Society (Lawrence and Livingston, Phys. Rev. **37**, 1707, (1931)) results of a preliminary study of the practicability of the method were given. Further work was reported in a Letter to the Editor of the Physical Review (Lawrence and Livingston, Phys. Rev. **38**, 834 (1931).

ion travels from the interior of one tube to the interior of the next there is always an accelerating field, and the final velocity of the ion on emergence from the system corresponds approximately to a voltage as many times greater than the applied voltage between adjacent tubes as there are tubes. The method is most conveniently used for the acceleration of heavy ions; for light ions travel faster and hence require longer systems of tubes for any given frequency of applied oscillations.

The present experimental method makes use of the same principle of repeated acceleration of the ions by a similar sort of resonance with an oscillating electric field, but has overcome the difficulty of the cumbersomely long accelerating system by causing, with the aid of a magnetic field, the ions to circulate back and forth from the interior of one electrode to the interior of another.

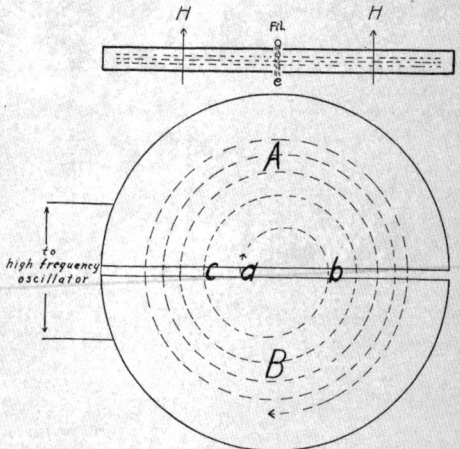

Fig. 1. Diagram of experimental method for multiple acceleration of ions.

This may be seen most readily by an outline of the experimental arrangement (Fig. 1). Two electrodes A, B in the form of semi-circular hollow plates are mounted in a vacuum tube in coplanar fashion with their diametral edges adjacent. By placing the system between the poles of a magnet, a magnetic field is introduced that is normal to the plane of the plates. High frequency electric oscillations are applied to the plates so that there results an oscillating electric field in the diametral region between them.

With this arrangement it is evident that, if at one moment there is an ion in the region between the electrodes, and electrode A is negative with respect to electrode B, then the ion will be accelerated to the interior of the former. Within the electrode the ion traverses a circular path because of the magnetic field, and ultimately emerges again between the electrodes; this is indicated in the diagram by the arc a .. b. If the time consumed by the ion in making the semi-circular path is equal to the half period of the electric oscillations, the electric field will have reversed and the ion will receive a second acceleration, passing into the interior of electrode B with a higher velocity. Again it travels on a semi-circular path ($b..c$), but this time the radius of curvature is greater because of the greater velocity. For all velocities (neglecting variation of mass with velocity) the radius of the path is proportional to the velocity, so that the time required for traversal of a semi-circular path is independent of the ion's velocity. Therefore, if the ion travels its first half circle in a half cycle of the oscillations, it will do likewise on all succeeding paths. Hence it will circulate around on ever widening semi-circles from the interior of one electrode to the interior of the other, gaining an increment of energy on each crossing of the diametral region that corresponds to the momentary potential difference between the electrodes. Thus, if, as was done in the present experiments, high frequency oscillations having peak values of 4000 volts are applied to the electrodes, and protons are caused to spiral around in this way 150 times, they will receive 300 increments of energy, acquiring thereby a speed corresponding to 1,200,000 volts.

It is well to recapitulate these remarks in quantitative fashion. Along the circular paths within the electrodes the centrifugal force of an ion is balanced by the magnetic force on it, i.e., in customary notation,

$$\frac{mv^2}{r} = \frac{Hev}{c}. \tag{1}$$

It follows that the time for traversal of a semi-circular path is

$$t = \frac{\pi r}{v} = \frac{\pi mc}{He} \tag{2}$$

which is independent of the radius r of the path and the velocity v of the ion. The particle of mass m and charge e thus may be caused to travel in phase with the oscillating electric field by suitable adjustment of the magnetic field H: the relation between the wave-length λ of the oscillations and the corresponding synchronizing magnetic field H is in consequence

$$\lambda = \frac{2\pi mc^2}{He}. \tag{3}$$

Thus for protons and a magnetic field of 10,000 gauss the corresponding wave-length is 19.4 meters; for heavier particles the proper wave-length is proportionately longer.*

It is easily shown also that the energy V in volt-electrons of the charged particles arriving at the periphery of the apparatus on a circle of radius r is

---

* It should be mentioned that, for a given wave-length, the ions resonate with the oscillations when magnetic fields of 1/3, 1/5, etc., of that given by Eq. (3) are used. Such types of resonance were observed in the earlier experimental studies. In the present experiments, however, the high speed ions resulting from the primary type of resonance only were able to pass through the slit system to the collector, because of the high deflecting voltages used.

$$V = 150 \frac{H^2 r^2}{c^2} \frac{e}{m}. \tag{4}$$

Thus, the theoretical maximum producible energy varies as the square of the radius and the square of the magnetic field.

EXPERIMENTAL ARRANGEMENT

The experimental arrangement is shown diagrammatically in some detail in Fig. 2. Fig. 3 is a photograph of the brass vacuum tube with cover removed showing the filament, the accelerating electrode, the deflecting plates and slit system, the probe in front of the first slit mounted on a ground joint and the Faraday collector behind the last slit. An external view of the apparatus is shown in Fig. 4. Here the tube is shown between the magnet pole faces, connected with the oscillator, the vacuum system and hydrogen generator. This gives a good general idea of the modest extent of the equipment involved for the generation of protons having energies somewhat in excess of 1,000,000 volt-electrons. The control panel and electrometer, being on the other side, are not shown in the picture. The description of the apparatus follows.

The accelerating system. Though there are obvious advantages in applying the high frequency potentials with respect to ground to both accelerating electrodes, in the present experiments it was found convenient to apply the high frequency voltage to only one of the electrodes, as indicated in Fig. 2. This electrode was a semi-circular hollow brass plate 24 cm in diameter and 1 cm thick. The sides of the hollow plate were of thin brass so that the interior of the plate had approximately these dimensions. It was mounted on a water-cooled copper re-entrant tube which in turn passed through a copper to glass seal. The electrode insulated in this way was mounted in an evacuated brass box having internal dimensions 2.6 cm by 28.6 cm by 28.6 cm, there being thus a lateral clearance between the electrode and walls of the brass chamber of 8 mm.

The brass box itself constituted the other electrode of the accelerating system. Across the mid-section of the brass chamber parallel to the diametral edge of the electrode A was placed a brass dividing wall S with slits of the same dimensions as the opening of the nearby electrode. This arrangement gave rise to the same type of oscillating electric fields as would have been produced had there been used two insulated semi-circular electrodes with their diametral edges adjacent and parallel.

The source of ions. An ideal source of ions is one that delivers to the diametral region between the electrodes large quantities of ions with low components of velocity normal to the plane of the accelerators. This requirement has most conveniently been met in the present experiments merely by having a filament placed above the diametral region from which a stream of electrons pass down along the magnetic lines of force, generating ions of gases in the tube. The ions so formed are pulled out sideways by the oscillating electric field. The electrons are not drawn out because of their very small radii of curvature in the magnetic field. Thus, the beam of electrons is col-

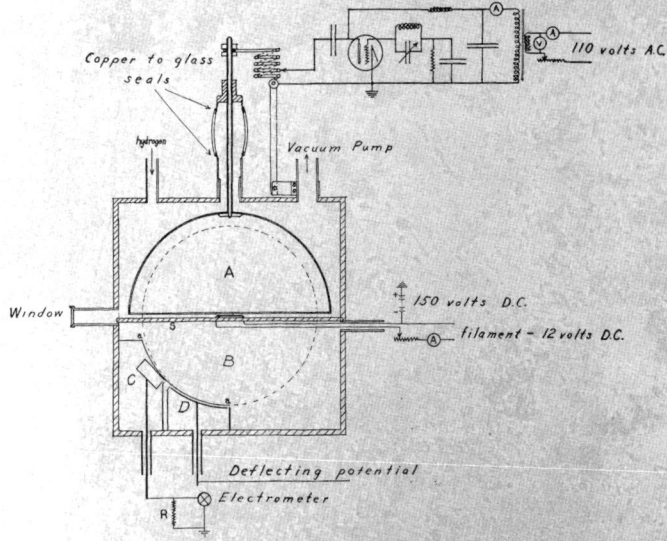

Fig. 2. Diagram of apparatus for the multiple acceleration of ions.

Fig. 3. Tube for the multiple acceleration of light ions—with cover removed.

limated and the ions are formed with negligible initial velocities right in the region where they are wanted. The oscillating electric field immediately draws them out and takes them on their spiral paths to the periphery. This arrangement is diagrammatically shown in the upper part of Fig. 1.

Fig. 4. External view of apparatus for generation of 1,220,000 volt protons.

The magnetic field. This experimental method requires a highly uniform magnetic field normal to the plane of the accelerating system. For example, if the ions are to circulate around 100 times, thereby gaining energy corresponding to 200 times the applied voltage, it is necessary that the magnetic field be uniform to a fraction of one percent. A general consideration of the matter leads one to the conclusion that, if possible, the magnetic field should be constant to about 0.1 percent from the center outward. Though this presumably difficult requirement has been met easily by an empirical method of field correction, the magnet used in the present experiments has pole faces machined as accurately as could be done conveniently. Its design was quite similar to that of Curtis.[18] The pole faces were 11 inches in diameter and the gap separation was $1\frac{1}{2}$ inches. Armco iron was used throughout the magnetic circuit. The magnetomotive force was provided by two coils of number 14 double cotton covered wire of 2,000 turns each. No water cooling was incorporated, for the magnet was not intended for high fields. In practice the magnet would give a field of 14,000 gauss for considerable periods without overheating. The pole faces were made parallel to about 0.2 percent and so it was to be expected that the magnetic field produced would be highly uniform. Exploration with a bismuth spiral confirmed this expectation, since it failed to show an appreciable variation of the magnetic field in the region between the poles, excepting within an inch of the periphery.

The collector system. In planning a suitable arrangement for collecting the high speed ions at the periphery of the apparatus, it was clearly desirable to devise something that would collect the high speed ions only and which would also measure their speeds. One might regard it as legitimate to suppose that the magnetic field itself and the distance of the collector from the center of the system would determine the speeds of the ions collected. This would be true provided there were no scattering and reflection of ions. To eliminate these extraneous effects a set of 1 mm slits was arranged on a circle $a \ldots a$, as shown in Fig. 2, of radius about 12 percent greater than the circle, indicated by the dotted line in the figure, having its center at the center of the tube and a radius of 11.5 cm. The two circles were tangent at the first slit as shown. The ions on arrival at the first slit would be traveling presumably on circles approximately like the dotted line, and hence would not be able to pass through the second and third slits to the Faraday collector C. Electrostatic deflecting plates D, separated by 2 mm, were placed between the first two slits, making possible the application of electrostatic fields to increase the radius of curvature of the paths of the high speed ions sufficiently to allow them to enter the collector. By applying suitable high potentials to the deflecting system in this way, only correspondingly high speed ions were registered.

The collector currents were measured by an electrometer shunted with a suitable high resistance leak.

The oscillator. The high frequency oscillations applied to the electrode were supplied by a 20 kilowatt Federal Telegraph water-cooled power tube in a "tuned plate tuned grid" circuit, for which the diagram of Fig. 2 is self-explanatory.

THE FOCUSSING ACTIONS

When one considers the circulation of the ions around many times as they are accelerated to high speeds in this way, one wonders whether in practice an appreciable fraction of those starting out can ever be made to

---

[18] L. F. Curtis, Jour. Op. Soc. Am. **13**, 73 (1926).

arrive at the periphery and to pass through a set of slits perhaps 1 mm wide and 1 cm long. The paths of the ions in the course of their acceleration would be several meters, and, because of the unavoidable spreading effects of space charge, thermal velocities and contact electromotive forces, as well as inhomogeneities of the applied fields, it would appear that the effective solid angle of the peripheral slit for the ions starting out would be exceedingly small.

Fortunately, however, this does not turn out to be the case. The electric and magnetic fields have been so arranged that they provide extremely strong focussing actions on the spiraling ions, which keep them circulating close to the median plane of the accelerating system.

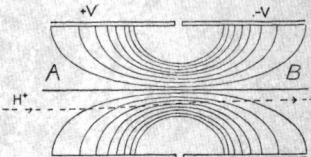

Fig. 5. Diagram indicating the focussing action of the electric field between the accelerating electrodes.

Fig. 5 shows the focussing action of the electric fields. There is depicted a cross-section of the diametral region between the accelerating electrodes with the nature of the field indicated by lines of force. There is shown also a dotted line which represents qualitatively the path of an ion as it passes from the interior of one electrode to the interior of the other. It is seen that, since it is off the median plane in electrode A, on crossing to B it receives an inward displacement towards the median plane. This is because of the existence of the curvature of the field, which over certain regions has an appreciable component normal to the plane, as indicated. If the velocity of the ion is very

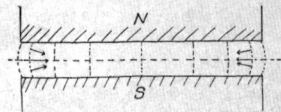

Fig. 6. Diagram indicating focussing action of magnetic field.

high in comparison to the increment of velocity gained in going from plate A to plate B, its displacement towards the center will be relatively small and, to the first approximation, it may be described as due to the ion having been accelerated inward on the first half of its path across and accelerated outward by an equal amount during the remainder of its journey, the net result being a displacement of the ion towards the center without acquiring a net transverse component of velocity. In general, however, the outward acceleration during the second half will not quite compensate the inward acceleration of the first, resulting in a gain of an inward component of velocity as well as an inward displacement. In any event, as the ion spirals around it will migrate back and forth across the median plane and will not be lost to the walls of the tube.

The magnetic field also has a focussing action. Fig. 6 shows diagrammatically the form of the field produced by the magnet. In the central region of the pole faces the magnetic field is quite uniform and normal to the plane of the faces; but out near the periphery the field has a curvature. Ions traveling on circles near the periphery experience thereby magnetic forces, indicated by the arrows. If the circular path is on the median plane then the magnetic force is towards the center in that plane. If the ion is traveling in a circle off the median plane, then there is a component of magnetic force that accelerates it towards the median plane, thereby giving effectively a focussing action.

We have experimentally examined these two focussing actions, using a probe in front of the first slit of the collector system that could be moved up and down across the beam by means of a ground joint (see Fig. 3). It was

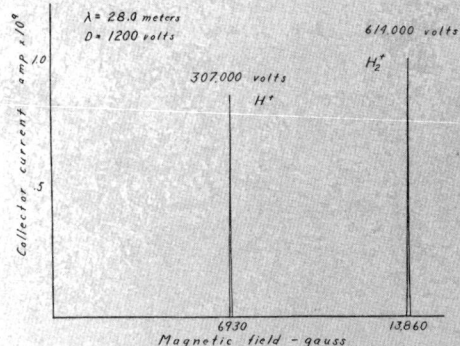

Fig. 7. Ion current to Faraday collector as a function of the magnetic field with oscillations of 28 meters wave-length applied to the accelerating electrodes.

found that the focussing actions were so powerful that *the beam of high speed ions had a width of less than one millimeter.* Such a narrow beam of ions of course is ideal for many experimental studies.

As a further test of the focussing action of the two fields, the median plane of the accelerating system was lowered 3 mm with respect to the plane of symmetry of the magnetic field. It was found that the high speed ion beam at the periphery traveled in a plane that was between the planes of symmetry of the two fields showing that both focussing actions were operative and at the periphery were of the same order of magnitude.

Experimental Results

As a typical example there is shown in Fig. 7 a plot of the ion current to the Faraday collector as a function of the magnetic field for applied oscillations of wave-length 28 meters and with hydrogen in the tube. It is seen that there are only two narrow ranges of magnetic field strength over which ion currents are observed; both correspond exactly to expectations, the one at

6930 gauss involving the resonance of protons, the other, hydrogen molecule ions.

For each wave-length used, *the magnetic field giving the greatest current to the collector agreed precisely with the theoretically expected value.* This is illustrated in Fig. 8 where the curves represent the theoretical hyperbolic relations between wave-length and magnetic field (Eq. 3) for protons and hydrogen molecule ions, and the circles represent the experimental observations. The magnetic fields were measured with a bismuth spiral and the oscillation wave-lengths were determined with a General Radio wavemeter. No effort was made to obtain considerable precision in these measurements, and in consequence their accuracy was hardly greater than 1 percent.

The variation with applied high frequency voltage of the widths of the resonance peaks agreed also with theoretical expectations. It was found that as the voltage was reduced the peaks became sharper, and indeed, with voltages such that the ions were required to spiral around fifty and more times to reach

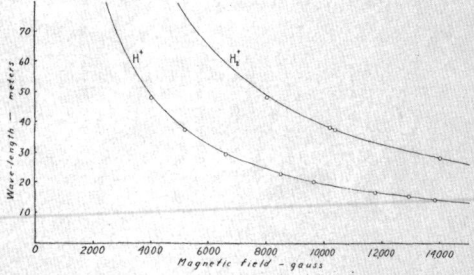

Fig. 8. Magnetic fields producing resonance of ions with oscillations of various wave-lengths: the curves are the theoretical relations (Eq. (3)) for H^+ and H_2^+ ions and the circles are the experimental observations.

the periphery, the ion currents diminished practically to zero when the magnetic field was changed a few tenths of one percent from the optimum value. This sharpness of resonance is understandable when it is remembered that the time required for an ion to execute one of its semi-circular paths is inversely proportional to the magnetic field. If, for example, the magnetic field were one percent greater or less than the resonance value, the ions would find themselves completely out of phase with the oscillations after having made fifty revolutions in the tube. In Fig. 7 the peaks exhibit an appreciable width, and indeed they extend over a one percent range of magnetic field. In most of the experiments, however, the ions circulated around many more times resulting in peaks of such restricted breadth as scarcely to be discernible in a diagram of this sort.

It is of course evident that the upper limit to the number of times the ions will circulate is determined by the degree of uniformity of the average value of the magnetic field along the spiral paths. Indeed, it would seem difficult to construct a magnet with pole faces giving fields of sufficient uniformity to allow more than 100 accelerations of the ions. But happily there is a very simple empirical way of correcting for the lack of uniformity of the field, that makes possible a surprisingly large voltage amplification. This is accomplished by insertion of thin sheets of iron between the tube and the magnet; either in the central region or out towards the periphery, as may be needed. If the magnetic field is, on the average, slightly less out towards the periphery so that the ions lag in phase more and more with respect to the oscillations as they spiral around, they may be brought back into step again by the insertion near the periphery of a strip of iron of suitable width, thickness and extension. If, on the other hand, the ions tend to get ahead in phase in this region, an effective correction can be made by inserting a suitable iron sheet in the central region.

It should be emphasized in this connection that the requirement is not that the magnetic field has to be uniform everywhere to the extent indicated above; small deviations from uniformity are allowable provided that the average value of the magnetic field over the paths of the ions is such that they traverse successive revolutions in equal intervals of time. Thus, small magnetic field adjustments can be accomplished by increasing or decreasing the field over small portions of successive circular paths of the ions. In the present experiments the most satisfactory adjustment was made by the insertion of a sheet of iron 0.025 cm thick having a shape much like an exclamation point extending radially with the thick end 8 cm wide in the central region and the narrow end 3 cms wide at the periphery. Insertion of this correcting "shim" *increased the amplification factor* (that is, the ratio of the equivalent voltage of the ions arriving at the collector to the maximum high frequency voltage applied to the tube) *from about 75 to about 300*. These figures are of necessity somewhat rough estimates, because no means were conveniently at hand to measure the high frequency voltages applied to the tube. Our estimates are based solely on sparking distances in air, and hence it is not unlikely that the voltage amplifications were even greater.

The greatest voltage amplification was obtained when generating the highest speed ions, 1,220,000 volt-protons. In all our work we have found the experimental method to be increasingly effective in this regard, as in others, as we go to higher voltages.

For example, the optimum pressure of hydrogen in the tube has been found to increase from less than 10^{-4} mm of Hg when generating 200,000 volt-protons to more than 10^{-3} mm when producing 1,000,000 volt-protons. By the optimum pressure is meant the pressure that gives the largest current to the collector for a given electron emission from the filament. The reason for this is, of course, connected with the fact that the effective mean free path of the spiralling particles increases with voltage.

Examples of the observed variation with voltage on the deflecting plates of the ion currents to the collector are shown in Fig. 9. Each curve is for a particular resonance condition; curve A, for example, was obtained when protons resonated with 37.5 meter oscillations in a magnetic field of 5180 gauss, thereby theoretically resulting in the arrival of 172,000 volt-protons at the first slit of the collector system. The wave-lengths used and the theoretically expected equivalent voltages of the ions generated in each instance is indicated in the figure. It is seen that, the higher the equivalent voltage of the ions, the higher was the required deflecting voltage to obtain the maximum ion currents to the collector. Indeed, within the experimental error, the optimum deflecting voltage was proportional to the theoretical kinetic energies of the ions (calculated from Eq. (4)) and was quite independent of the magnitude of the high frequency voltage applied to the accelerating electrode. *These observations constitute incontrovertible evidence that the ions arriving at the collector actually had the high speeds theoretically expected. The observed absolute magnitudes of the deflecting voltages also agreed with theoretical calculations* within the experimental uncertainty of the paths of the ions before entering the deflecting system. Because of the considerable width of the ion source (the filament was 2.5 cm long) the effective center of

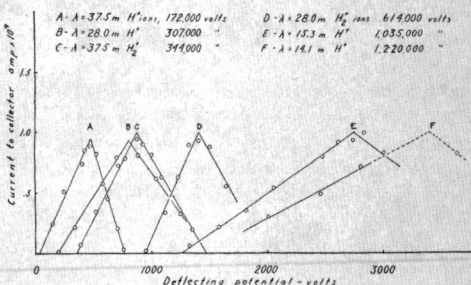

Fig. 9. Ion currents to the Faraday collector as a function of the voltage applied to the deflecting plates. The optimum deflecting voltages are seen to be proportional to the theoretically calculated kinetic energies of the ions (indicated in the figure in volts), thus proving that the ions arriving at the collector actually have the theoretically expected high speeds.

the circular paths of the ions at the periphery was quite broad. This fact together with the slit widths accounted for the absolute range of deflecting voltages over which ion currents reached the collector.

Discussion

The present experiments have accomplished one of the objectives set forth in the introduction, namely, the development of a convenient method for the production of protons having kinetic energies of the order of magnitude of 1,000,000 volt-electrons. It is well to emphasize two particular features that have contributed more than anything else to the effectiveness of the method: the *focussing actions of the electric and magnetic fields*, and the *simple means of empirically correcting the magnetic field* by the introduction of suitable iron strips. The former has solved the practical problem of generation of intense high speed ion beams of restricted cross-section so much desired in studies of collision processes. The latter has eliminated the problem of uniformity of magnetic field, making possible voltage amplifications of more than 300. This in turn has practically eliminated any difficulties associated with generation and application to the accelerating electrodes of required high frequency voltages. In consequence, we have here a source of high speed light ions that is readily constructed and assembled in a relatively small laboratory space out of quite modest laboratory equipment. The beam of ions so produced has valuable characteristics of convenience and flexibility for many experimental investigations; there are obvious advantages of a steady beam of high speed ions of but one millimeter diameter generated in an apparatus on an ordinary laboratory table. Moreover, the apparatus evolved in the present work is in no respects capricious, but functions always in a satisfactorily predictable fashion. This is illustrated by the fact that the accelerating tube can be taken apart and reassembled, and then within a few hours after re-evacuation steady beams of 1,200,000 protons can always be obtained.

But it is perhaps of even more interest to inquire as to the practical limitations of the method; to see what extensions and developments are foreshadowed by the present experiments.

Of primary importance is the probable experimental limitation on the producible proton energies. The practical limit is set by the size of the electromagnet available; for the final equivalent voltage of the ions at the periphery is proportional to the square of the magnetic field strength and to the square of the radius of the path. For protons, it is not feasible to use magnetic fields much greater than employed in the present work (about 14,000 gauss) because of the difficulties of application of suitably higher frequency oscillations—that is to say, it is not desirable to go much below 14 meters wavelength. However, it is entirely practicable to use a much larger magnet than that employed in the present experiments. At the present time a magnet having pole faces 114 cm in diameter is being installed in our laboratory. As will be seen from Eq. (4), a magnetic field of 14,000 gauss over such a large region *makes possible the production of 25,000,000 volt-protons.*

Of course, it may be argued that there are other difficulties which preclude ever reaching such a range of energies. For example, there is the question of whether it is possible to obtain such a great amplification factor that the high frequency voltages necessarily applied to the accelerating electrodes are low enough to be realizable in practice. In the present experiments an amplification of 300 was obtained with no great effort, and it would seem that with more careful correction of the field this amplification could be considerably increased at higher voltages. In the higher range of speeds the variation of mass with velocity begins to be appreciable, but presents no difficulty as it can be allowed for by suitable alteration of the magnetic field in the same empirical manner as is done to correct its otherwise lack of uniformity.

Assuming then a voltage amplification of 500, the production of 25,000,-000 volt-protons would require 50,000 volts at a wave-length of 14 meters applied across the accelerators; thus, 25,000 volts on each accelerator with respect to ground. *It does appear entirely feasible to do this*, although to be sure a considerable amount of power would have to be supplied because of the capacity of the system.

Of similar interest is the matter of maximum obtainable beam intensities. In the present experiments no efforts have been made to obtain high intensities and the collector currents have usually been of the order of magnitude of 10^{-9} amp. Using the present method of generation of the ions, there are two factors that can be drawn upon to increase the yield of high speed ions—the electron emission and the pressure of hydrogen in the tube. The electron emission can easily be increased from 10 to 100 times over that used in the present experiments. The effective free paths of the protons increase with voltage so that, as was found to be the case, the maximum usable pressure of hydrogen is governed by the setting in of a high frequency discharge in the tubes due to the voltage on the accelerators. This appears to occur at a pressure greater than 10^{-3} mm of Hg; the reason the critical pressure is so high is probably to be associated with the quenching action of the magnetic field. These considerations make it seem reasonable to expect that, using the present ion source, *high speed ion currents of as much as 0.1 microampere can readily be obtained.*

At all events, it seems that the focussing of the spiralling ions is so effective that a quite considerable portion of those starting out arrive at the collector and that the beam intensity is determined largely by the source. *This method of multiple acceleration is capable of yields of the same order of magnitude as would conceivably result from the direct application of high voltages.*

For a given experimental arrangement the energy of the ions arriving at the collector varies inversely as their masses and directly as their charges. Thus, the large magnet mentioned above makes possible the production of 12,500,000 volt hydrogen molecule ions and doubly charged helium ions (alpha-particles) as well as 25,000,000 volt-protons. Moreover, generating the theoretically maximum value of ion energies becomes much easier with increasing atomic weight because the wave-length of the applied high frequency oscillations increases in a like ratio. For example, using a magnetic field of 14,000 gauss over a region 114 cm in diameter, 2,800,000 volt nitrogen ions could be generated by applying 123 meter oscillations. Broadly speaking, then, the apparatus is well adapted to the production of ions of all the elements up to atomic weight 25 having kinetic energies in excess of 1,000,000 volt-electrons.

We wish to express our gratitude and thanks to the Committee-on-Grants-in-Aid of the National Research Council, the Federal Telegraph Company through the courtesy of Dr. Leonard F. Fuller, Vice-President, the Research Corporation, and the Chemical Foundation for their generous assistance which has made these experiments possible.

Fig. 3. Tube for the multiple acceleration of light ions—with cover removed.

Fig. 4. External view of apparatus for generation of 1,220,000 volt protons.

논문 웹페이지

Charge Independence for V-particles*

Tadao Nakano and Kazuhiko Nishijima

Department of Physics, Osaka City University

November 16, 1953

Assuming the charge independence for V-particles, the qualitative features of these unstable heavy particles are investigated.

In view of the present experimental material, there seems to be three charge states for V_1:

(1) V_1°: This particle has been most thoroughly investigated by many workers, and known to decay as $V_1^\circ \to p + \pi^- + Q$, $Q \sim 37$ Mev.

(2) V_1^+: This particle was discovered by the Pasadena group.[1] $V_1^+ \to p + \pi^\circ + Q$, $Q \sim 40$ Mev.

(3) V_1^-: One case was found in the cosmotron experiments[2] that seems to require the existence of V_1^-, although not conclusive. It is as yet not clear whether the isotopic spin of V_1 is 1/2 or 1 or higher. We shall, however, tentatively assign it as equal to 1, since this case is of special interest. Then from the cosmotron experiments,[3] V_4° or V_2° which is tentatively denoted as Π° should have a half integral isotopic spin[4] in reference to the process

$$\pi^- + p \to V_1^\circ + \Pi^\circ, \quad (\Pi^\circ \to \pi^+ + \pi^-). \quad (1)$$

If we assume that there is no doubly charged counter particle to Π°, the isotopic spin of Π should be 1/2. In such a case Π^+ and Π° are treated just as proton and neutron so long as we are concerned with their transformation properties in isotopic space. Hence the Π°-particle should be described by a complex wave function as well as the charged Π-particle, and we must distinguish between the Π°-particle and its anti-particle $\widetilde{\Pi}^\circ$. This distinction leads to many interesting results as we shall see later.

From the above isotopic spin assignment we have the following results.

(1) The "even-odd" rule[5] is an inevitable consequence of the charge independence. If both the spin and isotopic spin of a hot particle* are integer or half-integer we call it an even particle, whereas if only one of them is integer and the other is half-integer we call it an odd particle. The even-odd rule holds for such an even-odd assignment of hot particles. Hence the large abundance and the striking stability of the V-particles against π- or γ-decay are automatically guaranteed. Recently Pais derived this rule from his own theory of the "ω"-space[6] by imposing the conservation of the ω-parity, while in the present work it is derived with less new elements.

(2) In production processes, we have the following conservation law valid for the charge independent and electromagnetic interactions

$$n(V_1) - n(\Pi) = \text{const.}, \quad (2)$$

where $n(V_1)$ is the no. of V_1-particles minus the no. of anti-V_1-particles and $n(\Pi)$ the no. of Π^+ and Π° minus the no. of $\widetilde{\Pi}^-$ and $\widetilde{\Pi}^\circ$. This law is proved as follows.

From the above isotopic spin assignment for V_1- and Π-particles, we have

$$q = I_3 + 1/2(n(N) + n(\Pi)), \quad (3)$$

where q and I_3 are the total charge and the third component of the isotopic spin of the system of hot particles.

There is another conservation law, the conservation of baryons**

$$b = n(V_1) + n(N) = \text{const.}. \quad (4)$$

Since q, b and I_3 are conserved for the charge independent and electro-magnetic interactions, we have from (3) and (4)

$$n(V_1) - n(\Pi) = b - 2(q - I_3) = \text{const.}.$$

---

* After the completion of this work, the authors knew in a private letter from Prof. Nambu to Prof. Hayakawa that Dr. Gell-Mann has also developed a similar theory.

* By a "hot particle", we mean a particle with strong nuclear interaction.

** The "baryon" is the collective name for the members of the nucleon family. This name is due to Pais. See ref. (6).

Especially in pion-nucleus or nucleon-nucleus impacts, (2) can be written as

$$n(V_1)=n(\Pi)=n(\Pi^+,\Pi^\circ)-n(\widetilde{\Pi}^-,\widetilde{\Pi}^\circ). \quad (5)$$

It must be noticed that in cases of heavy nuclei the Coulomb effects cannot be discarded and hence the validity of the conservation law for the electromagnetic interaction is necessary.

The conservation law (5) can forbid many processes, e.g.

$$\pi^- + p \to V_1 + \widetilde{\Pi}^-, \quad (6)$$
$$N+N \to V_1+V_1, \text{ etc. } (N:\text{nucleon}). \quad (7)$$

Hence the production of V-particles in nucleon-nucleon collisions will be due to the processes such as

$$N+N \to N+V_1+\Pi, \quad (8)$$
or $\quad N+N \to N+N+\Pi+\widetilde{\Pi}. \quad (9)$

(3) Since the process (7) is forbidden, we may conjecture that the production of V-particles will result mainly in pion-nucleon (or nucleus) collisions rather than in nucleon-nucleon (or nucleus) collisions at energies where the cosmic ray experiments are being performed, i.e. at about 10 Mev in the laboratory system and about 2 Mev in the centre of mass system, in conformity with the experimental viewpoint.[7]

(4) From the rarity of the σ-stars produced by negative heavy mesons the Bristol group[8] conjectured that the positive K-particles might be much more abundant than the negative ones.

Since the V_1-particles are supposed to be more easily produced than the Π- mesons because of the lower excitation energy to transform a nucleon into a V_1-particle than the energy required to create a heavy Π-meson, the production of V_1-particles in high energy nuclear events will occur in such a manner that $n(V_1)$ assumes as large a value as possible for a fixed value of $n(\Pi^+,\Pi^\circ)+n(\widetilde{\Pi}^-,\widetilde{\Pi}^\circ)+n(V_1)$. Then from eq. (5), we may expect

$$n(V_1) \sim n(\Pi^+,\Pi^\circ) \gg n(\widetilde{\Pi}^-,\widetilde{\Pi}^\circ). \quad (10)$$

If we identify the K-particles with the Π-particles, then the conjecture of the Bristol group can be interpreted in terms of the relation (10). However, in the cloud chamber experiments both the positive and negative V-particles lighter than nucleon are observed comparably.[9] Thus it is an important problem to settle how many kinds of charged V-particles are present in nature.*

(5) The selection rules imposed by the charge conjugation and charge symmetry[10] cannot be applied to Π-meson decays. Since the real and imaginary parts of the complex wave function of Π° have opposite parities under charge conjugation, we cannot apply the selection rule

$$n(v)+n(t)=\text{odd is forbidden}, \quad (11)$$

for transition among neutral Bosons, to Π°.

For charged Π-particles the CT transformation[10] when applied to $\Pi^\pm$ alters its charge state, so that in this case, too, we cannot apply the second selection rule

$$n(v)+n(t)+n(\tau_3)=\text{odd is forbidden}, \quad (12)$$

for non-radiative transitions, to $\Pi^\pm$.

The authors thank Prof. S. Hayakawa for his kind criticism.

1) York, Leighton, and Bjornerud, Phys. Rev. 90 (1953), 167.
2) C. N. Yang's report on the cosmotron experiments. Lecture at the International Confererence on Theoretical Physics, held at Kyoto, Sept. 1953. Some V_1 decays seem to contradict the evenodd rule by exhibiting a cascade decay of a heavy unstable particle. See in this connection ref. (4) in which the generalization of the evenodd rule is discussed.
3) Fowler, Shutt, Thorndike, and Whittemore, Phys. Rev. 91 (1953), 1287. See also ref. (2).
4) K. Nishijima, Prog. Theor. Phys. 9 (1953), 414. In this paper, it was stated that a Boson cannot assume a half-integral isotopic spin. However, there are special exceptional cases which are the matter of interest in the present work.
5) A. Pais, Phys. Rev. 86 (1952), 663.
6) A. Pais, Lecture at the International Conference on Theoretical Physics. Prog. Theor. Phys. 10 (1953), 457.
7) Leighton, Wanlass, and Anderson, Phys. Rev. 89 (1953), 148.
8) See for instance, Fowler, Menon, Powell, and Rochat, Phil. Mag. 42 (1951), 1040.
9) Astbury et al., Phil. Mag. 43 (1952), 1283. Astbury et al., ibid. 44 (1953), 242.
10) A. Pais and R. Jost, Phys. Rev. 87 (1952), 871.

---

* If particles with different names correspond to the different modes of decay of an identical particle, the branching ratio of these modes should be constant independently of the mechanisms and energies of the production processes.

On a Composite Model for the New Particles*

Shoichi Sakata

Institute for Theoretical Physics, Nagoya University, Nagoya

September 3, 1956

Recently, Nishijima-Gell-Mann's rule[1] for the systematization of new particles has achieved a great success to account for various facts obtained from the experiments with cosmic rays and with high energy accelerators. Nevertheless, it would be desirable from the theoretical standpoint

---

* The content of this letter was read before the annual meeting of the Japanese Physical Society held in October 1955.

A note on the same supject has also been published in Bulletin de L'acadèmie Polonaise des Sciences (Cl, III-vol. IV, No. 6, 1956)

to find out a more profound meaning hidden behind this rule. The purpose of this work is concerned with this point.

It seems to me that the present state of the theory of new particles is very similar to that of the atomic nuclei 25 years ago. At that time, we had known a beautiful relation between the spin and the mass number of the atomic nuclei. Namely, the spin of the nucleus is always integer if the mass number is even, whereas the former is always half integer if the latter is odd. But unfortunately we could not understand the profound meaning for this even-odd rule. This fact together with other mysterious properties of the atomic nuclei, for instance the beta disintegration in which the conservation of energy seemd to be invalid, led us to a very pessimistic view-point that the quantum theory would not be applicable in the domain of the atomic nucleus. However the situation was entirely changed after the discovery of the neutron. Iwanenko and Heisenberg[2] proposed immediately a new model for the atomic nuclei in which neutrons and protons are considered to be their constituents. By assuming that the neutron has the spin of one half, they explained the even-odd rule for the spins of atomic nuclei as the result of the addition law for the angular momenta of the constituents. Moreover, they could reduce all the mysterious properties of atomic nuclei to those of the neutron contained in them.

Supposing that the similar situation is realized at present, I proposed a compound hypothesis for new unstable particles to account for Nishijima-Gell-Mann's rule. In our model, the new particles are considered to be composed of four kinds of fundamental particles in the true sense, that is, nucleon, antinucleon, Λ^0 and anti-Λ^0. If we assume that Λ^0 has such intrinsic properties as were assigned by Nishijima and Gell-Mann, we can easily get their even-odd rule for the composite particles as the result of the addition laws for the ordinary spin, the isotopic spin and the strangeness. In the next table, the models and the properties of the new particles are shown together with those of the fundamental particles in the true sense.

Name	Model	Isotopic Spin	Strangeness	Ordinary Spin
$\mathfrak{N}$		1/2	0	1/2
$\overline{\mathfrak{N}}$		1/2	0	1/2
Λ		0	−1	1/2 ?
$\overline{\Lambda}$		0	1	1/2 ?
π	$\mathfrak{N}+\overline{\mathfrak{N}}$	1	0	0
$\theta(\tau)$	$\mathfrak{N}+\overline{\Lambda}$	1/2	1	0 ?
$\bar{\theta}(\bar{\tau})$	$\overline{\mathfrak{N}}+\Lambda$	1/2	−1	0 ?
Σ	$\mathfrak{N}+\overline{\mathfrak{N}}+\Lambda$	1	−1	1/2 ?
Ξ	$\overline{\mathfrak{N}}+\Lambda+\Lambda$	1/2	−2	1/2 ?

Here $\mathfrak{N}$ and $\overline{\mathfrak{N}}$ denote nucleon and antinucleon respectively, whereas Λ and $\overline{\Lambda}$ denote Λ^0 and anti-Λ^0 respectively[3].

So far as the internal structure is not concerned, our model for new particles is identical with that of Nishijima and Gell-

Mann. However, it should be stressed that the curious properties of the new particles could be reduced to those of Λ^0, just like the mysterious properties of the atomic nuclei were reduced to those of neutron. Hence our theory contains less arbitrary elements than was the case for original one of Nishijima and Gell-Mann.

Though the rigorous treatment of our model is a very hard task[4], it is worthwhile to notice that most of the composite particles which seem to be stable against the strong interaction can be identified with the well-known new particles, and that there are possibilities of predicting some more new particles which have not been discovered up till now.[5]

Finally, it should be remarked that there are some other arguments in favour of the compound hypothesis for the elementary particles. In spite of the great success achieved by the advent of Tomonaga-Schwinger's technique, it has recently become clear that we could not avoid the internal inconsistency of the quantum field theory, so far as the point model for elementary particles was adopted. Moreover, in the case of π-meson, the cut-off prescription has recently been proved to be very powerful in order to account for the experimental results. These facts indicate strongly the necessity of substantial innovations in the model for the elementary particles, though some change has already been made by the discovery of the renormalization technique. Landau pointed out that the model for the electron would possibly be changed by the effect of the gravitational field. But in the case of π-meson we must look for another effect, because the cut-off radius is found to be as large as the order of the nucleon Compton wave length in contrast to $e^2/mc^2 \cdot e^{-137}$ $\sim 10^{-58}$ cm which appeared in the quantum electrodynamics.[6]

1) T. Nakano and K. Nishijima, Prog. Theor. Phys. **10** (1953), 581; K. Nishijima, Prog. Theor. Phys. **12** (1954), 107; **13** (1955), 285; M. Gell-Mann, Phys. Rev. **92** (1953), 833.
2) D. Iwanenko, Nature **129** (1932), 798; W. Heisenberg, ZS. Phys. **77** (1932), 1.
3) Markov (Rep. Acad. Sci. USSR, 1955) proposed also a composite model which is very similar to ours. It should be remarked that our model may be considered as a generalization of the π-meson model proposed by Fermi and Yang (Phys. Rev. **76** (1948), 1739), and that it will throw a new light on Heisenberg's theory of elementary particles (Zs. Naturf. **9a** (1954), 291; **10a** (1955), 425), in which only one kind of "Urmaterie" is assumed.
4) S. Tanaka, Prog. Theor. Phys. **16** (1956), 625. Z. Maki, Prog. Theor. Phys. **16** (1956), 667.
5) K. Matsumoto, Prog. Theor. Phys. **16** (1956), 583.
6) M. A. Markov, Uspekhi Fiz. Nauk **51** (1953), 317; L. Landau et al., DAN. **95** (1954), 497, 733, 1177; **96** (1954), 261; **102** (1955), 489; S. Kamefuchi & H. Umezawa, Prog. Theor. Phys. **15** (1956), 298; Nuovo Cimento **3** (1956), 1060.

논문 웹페이지

Symmetries of Baryons and Mesons*

MURRAY GELL-MANN
California Institute of Technology, Pasadena, California
(Received March 27, 1961; revised manuscript received September 20, 1961)

The system of strongly interacting particles is discussed, with electromagnetism, weak interactions, and gravitation considered as perturbations. The electric current j_α, the weak current J_α, and the gravitational tensor $\theta_{\alpha\beta}$ are all well-defined operators, with finite matrix elements obeying dispersion relations. To the extent that the dispersion relations for matrix elements of these operators between the vacuum and other states are highly convergent and dominated by contributions from intermediate one-meson states, we have relations like the Goldberger-Treiman formula and universality principles like that of Sakurai according to which the ρ meson is coupled approximately to the isotopic spin. Homogeneous linear dispersion relations, even without subtractions, do not suffice to fix the scale of these matrix elements; in particular, for the nonconserved currents, the renormalization factors cannot be calculated, and the universality of strength of the weak interactions is undefined. More information than just the dispersion relations must be supplied, for example, by field-theoretic models; we consider, in fact, the equal-time commutation relations of the various parts of j_4 and J_4. These nonlinear relations define an algebraic system (or a group) that underlies the structure of baryons and mesons. It is suggested that the group is in fact $U(3) \times U(3)$, exemplified by the symmetrical Sakata model. The Hamiltonian density θ_{44} is not completely invariant under the group; the noninvariant part transforms according to a particular representation of the group; it is possible that this information also is given correctly by the symmetrical Sakata model. Various exact relations among form factors follow from the algebraic structure. In addition, it may be worthwhile to consider the approximate situation in which the strangeness-changing vector currents are conserved and the Hamiltonian is invariant under $U(3)$; we refer to this limiting case as "unitary symmetry." In the limit, the baryons and mesons form degenerate supermultiplets, which break up into isotopic multiplets when the symmetry-breaking term in the Hamiltonian is "turned on." The mesons are expected to form unitary singlets and octets; each octet breaks up into a triplet, a singlet, and a pair of strange doublets. The known pseudoscalar and vector mesons fit this pattern if there exists also an isotopic singlet pseudoscalar meson χ^0. If we consider unitary symmetry in the abstract rather than in connection with a field theory, then we find, as an attractive alternative to the Sakata model, the scheme of Ne'eman and Gell-Mann, which we call the "eightfold way"; the baryons N, Λ, Σ, and Ξ form an octet, like the vector and pseudoscalar meson octets, in the limit of unitary symmetry. Although the violations of unitary symmetry must be quite large, there is some hope of relating certain violations to others. As an example of the methods advocated, we present a rough calculation of the rate of $K^+ \to \mu^+ + \nu$ in terms of that of $\pi^+ \to \mu^+ + \nu$.

I. INTRODUCTION

IN connection with the system of strongly interacting particles, there has been a great deal of discussion of possible approximate symmetries,[1] which would be violated by large effects but still have some physical consequences, such as approximate universality of meson couplings, approximate degeneracy of baryon or meson supermultiplets, and "partial conservation" of currents for the weak interactions.

In this article we shall try to clarify the meaning of such possible symmetries, for both strong and weak interactions. We shall show that a broken symmetry, even though it is badly violated, may give rise to certain exact relations among measurable quantities. Furthermore, we shall suggest a particular symmetry group as the one most likely to underlie the structure of the system of baryons and mesons.

We shall treat the strong interactions without approximation, but consider the electromagnetic, weak, and gravitational interactions only in first order.

The electromagnetic coupling is described by the matrix elements of the electromagnetic current operator $ej_\alpha(x)$. Likewise, the gravitational coupling is specified by the matrix elements of the stress-energy-momentum tensor $\theta_{\alpha\beta}(x)$, particularly the component $\theta_{44} = H$, the Hamiltonian density.

The weak interactions of baryons and mesons with leptons are assumed to be given (ignoring possible nonlocality) by the interaction term[2]

$$GJ_\alpha^\dagger J_\alpha^{(l)}/\sqrt{2} + \text{H.c.}, \qquad (1.1)$$

where the leptonic weak current $J_\alpha^{(l)}$ has the form

$$J_\alpha^{(l)} = i\bar{\nu}\gamma_\alpha(1+\gamma_5)e + i\bar{\nu}\gamma_\alpha(1+\gamma_5)\mu. \qquad (1.2)$$

We shall refer to $J_\alpha(x)$ as the weak current of baryons and mesons. Its matrix elements specify completely the weak interactions with leptons.

It is possible that the full weak interaction may be given simply by the term

$$G(J_\alpha + J_\alpha^{(l)})^\dagger (J_\alpha + J_\alpha^{(l)})/\sqrt{2}, \qquad (1.3)$$

although this form provides no explanation of the approximate rule $|\Delta I| = \tfrac{1}{2}$ in the nonleptonic decays of strange particles. If we can find no *dynamical* explanation of the predominance of the $|\Delta I| = \tfrac{1}{2}$ amplitude in these decays, we may be forced to assume that in addition to (1.3) there is a weak interaction involving the product

$$GL_\alpha^\dagger L_\alpha/\sqrt{2}, \qquad (1.4)$$

of charge-retention currents (presumably not involving leptons); or else we may be compelled to abandon (1.3)

---

* Research supported in part by U. S. Atomic Energy Commission and Alfred P. Sloan Foundation. A report of this work was presented at the La Jolla Conference on Strong and Weak Interactions, June, 1961.

[1] For example, see the "global symmetry" scheme of M. Gell-Mann, Phys. Rev. **106**, 1296 (1957) and J. Schwinger, Ann. Phys. **2**, 407 (1957).

[2] We use $\hbar = c = 1$. The Lorentz index α takes on the values 1, 2, 3, 4. For each value of α, the Dirac matrix γ_α is Hermitian; so is the matrix γ_5.

altogether. In any case, we shall define the weak current J_α by the coupling to leptons.

We shall assume microcausality and hence the validity of dispersion relations for the matrix elements of the various currents and densities. In addition, we shall sometimes require the special assumption of highly convergent dispersion relations.

Our description of the symmetry group for baryons and mesons is most conveniently given in the framework of standard field theory, where the Lagrangian density L of the strong interactions is expressed as a simple function of a certain number of local fields $\psi(x)$, which are supposed to correspond to the "elementary" baryons and mesons. Recently this type of formalism has come under criticism[3]; it is argued that perhaps none of the strongly interacting particles is specially distinguished as "elementary," that the strong interactions can be adequately described by the analyticity properties of the S matrix, and that the apparatus of field theory may be a misleading encumbrance.

Even if the criticism is justified, the field operators $j_\alpha(x)$, $\theta_{\alpha\beta}(x)$, and $J_\alpha(x)$ may still be well defined (by all their matrix elements, including analytic continuations thereof) and measurable in principle by interactions with external electromagnetic or gravitational fields or with lepton pairs. Since the Hamiltonian density H is a component of $\theta_{\alpha\beta}$, it can be a physically sensible quantity.

In order to make our description of the symmetry group independent of the possibly doubtful details of field theory, we shall phrase it ultimately in terms of the properties of the operators H, j_α, and J_α. In introducing the description, however, we shall make use of field-theoretic models. Moreover, in describing the behavior of a particular group, we shall refer extensively to a special example, the symmetrical Sakata model of Ohnuki et al.,[4] Yamaguchi,[5] and Wess.[6]

The order of presentation is as follows: We treat first the hypothesis of highly convergent dispersion relations for the matrix elements of currents; and we show that the notion of a meson being coupled "universally" or coupled to a particular current or density means simply that the meson state dominates the dispersion relations for that current or density at low momenta. Next we discuss the universality of strength of the currents themselves; evidently it cannot be derived from homogeneous linear dispersion relations for the matrix elements of the currents. We show that equal-time commutation relations for the currents fulfill this need (or most of it), and that, in a wide class of model field theories, these commutation rules are simple and reflect the existence of a symmetry group, which underlies the structure of the baryon-meson system even though some of the symmetries are badly violated. We present the group properties in an abstract way that does not involve the details of field theory.

Next, it is asked what group is actually involved. The simplest one consistent with known phenomena is the one suggested. It is introduced, for clarity, in connection with a particular field theory, the symmetrical Sakata model, in which baryons and mesons are built up of fundamental objects with the properties of n, p, and Λ. For still greater simplicity, we discuss first the case in which Λ is absent.

We then return to the question of broken symmetry in the strong interactions and show how some of the symmetries in the group, if they are not too badly violated, would reveal themselves in approximately degenerate supermultiplets. In particular, there should be "octets" of mesons, each consisting of an isotopic triplet with $S=0$, a pair of doublets with $S=\pm1$, and a singlet with $S=0$. In the case of pseudoscalar mesons, we know of π, K, and $\bar{K}$; these should be accompanied by a singlet pseudoscalar meson χ^0, which would decay into 2γ, $\pi^++\pi^-+\gamma$, or 4π, depending on its mass.

In Sec. VIII, we propose, as an alternative to the symmetrical Sakata model, another scheme with the same group, which we call the "eightfold way." Here the baryons, as well as mesons, can form octets and singlets, and the baryons N, Λ, Σ, and Ξ are supposed to constitute an approximately degenerate octet.

In Sec. IX, some topics are suggested for further investigation, including the possibility of high energy limits in which non-conserved quantities become conserved, and we give, as an example of methods suggested here, an approximate calculation of the rate of $K^+ \to \mu^+ + \nu$ decay from that of $\pi^+ \to \mu^+ + \nu$ decay.

II. MESONS AND CURRENTS

To introduce the connection between meson states and currents or densities, let us review the derivation[7] of the Goldberger-Treiman relation[8] among the charged pion decay amplitude, the strength of the axial vector weak interaction in the β decay of the nucleon, and the pion-nucleon coupling constant.

The axial vector term in J_α with $\Delta S=0$, $|\Delta I|=1$, $GP=-1$, can be written as $P_{1\alpha}+iP_{2\alpha}$, where $\mathbf{P}_\alpha$ is an axial vector current that transforms like an isotopic vector. We have, for nucleon β decay,

$$\langle N|\mathbf{P}_\alpha|N\rangle = u_f[i\gamma_\alpha F_{\mathrm{ax}}(s)+k_\alpha\beta(s)]\gamma_5(\tau/2)u_i, \quad (2.1)$$

where u_i and u_f are the initial and final spinors, k_α is the four-momentum transfer, and $s=-k^2=-k_\alpha k_\alpha$. At

---

[3] G. F. Chew, Talk at La Jolla Conference on Strong and Weak Interactions, June, 1961 (unpublished).
[4] M. Ikeda, S. Ogawa, and Y. Ohnuki, Progr. Theoret. Phys. (Kyoto) 22, 715 (1959); Y. Ohnuki, Proceedings of the 1960 Annual International Conference on High-Energy Physics at Rochester (Interscience Publishers, Inc., New York, 1960).
[5] Y. Yamaguchi, Progr. Theoret. Phys. (Kyoto) Suppl. No. 11, 1 (1959).
[6] J. Wess, Nuovo cimento 10, 15 (1960).

[7] J. Bernstein, S. Fubini, M. Gell-Mann, and W. Thirring, Nuovo cimento 17, 757 (1960). See also Y. Nambu, Phys. Rev. Letters 4, 380 (1960); and Chou Kuang-Chao, Soviet Phys.—JETP 12, 107 (1961).
[8] M. Goldberger and S. Treiman, Phys. Rev. 110, 1478 (1958).

$s=0$ we have just

$$F_{ax}(0) = -G_A/G, \qquad (2.2)$$

the axial vector renormalization constant.

The axial vector current is not conserved; its divergence $\partial_\alpha P_\alpha$ has the same quantum numbers as the pion ($J=0^-$, $I=1$). Between nucleon states we have

$$\langle N|\partial_\alpha \mathbf{P}_\alpha|N\rangle = \bar{u}_f i\gamma_5(\tau/2)u_i[2m_N F_{ax}(s) + s\beta(s)]. \quad (2.3)$$

We may compare this matrix element with that between the vacuum and a one-pion state

$$\langle 0|\partial_\alpha \mathbf{P}_\alpha|\pi\rangle = m_\pi^2 (2f_\pi)^{-1}\phi, \qquad (2.4)$$

where ϕ is the pion wave function and the constant f_π (or at least its square) may be measured by the rate of $\pi^+ \to \mu^+ + \nu$:

$$\Gamma_\pi = G^2 m_\pi m_\mu^2 (1 - m_\mu^2/m_\pi^2)^2 (f_\pi^2/4\pi)^{-1} (64\pi^2)^{-1}. \quad (2.5)$$

It is known that the matrix element (2.3) has a pole at $s = m_\pi^2$ corresponding to the virtual emission of a pion that undergoes leptonic decay. The strength of the pole is given by the product of m_π^2/f_π and the pion-nucleon coupling constant $g_{NN\pi}$. If we assume that the expression in brackets vanishes at large s, we have an unsubtracted dispersion relation for it consisting of the pole term and a branch line beginning at $(3m_\pi)^2$, the next lowest mass that can be virtually emitted:

$$2m_N F_{ax}(s) + s\beta(s) = (g_{NN\pi}/f_\pi) m_\pi^2 (m_\pi^2 - s)^{-1}$$
$$+ \int \sigma_{ax}(M^2) M^2 dM^2 \, (M^2 - s - i\epsilon)^{-1}. \quad (2.6)$$

At $s=0$, we have, using (2.2), the sum rule

$$2m_N(-G_A/G) = g_{NN\pi}/f_\pi + \int \sigma_{ax}(M^2)dM^2. \quad (2.7)$$

Now if the dispersion relation (2.6) is not only convergent but dominated at low s by the term with the lowest mass, then we have the approximate Goldberger-Treiman relation

$$2m_N(-G_A/G) \approx g_{NN\pi}/f_\pi, \qquad (2.8)$$

which agrees with experiment to within a few percent.

The success of the relation suggests that other matrix elements of $\partial_\alpha \mathbf{P}_\alpha$ may also obey unsubtracted dispersion relations dominated at low s by the one-pion term. For example, if we consider the matrix element between Λ and Σ, we should arrive at the relation

$$(m_\Lambda + m_\Sigma)(-G_A^{\Lambda\Sigma}/G) \approx g_{\Lambda\Sigma\pi}/f_\pi, \qquad (2.9)$$

if Λ and Σ have the same parity, or an analogous relation if they have opposite parity.

If such a situation actually obtains, then it may be said that the pion is, to a good approximation, coupled "universally" to the divergence of the axial vector current. To calculate any g approximately, we multiply the universal constant f_π, the sum of the initial and final masses, and the renormalization factor for the axial vector current.

Now let us turn to the case of a current that is conserved, say the isotopic spin current $\mathfrak{J}_\alpha$ with quantum numbers $J=1^-$, $I=1$. Acting on the vacuum, the operator $\mathfrak{J}_\alpha$ does not lead to any stable one-meson state, but it does lead to the unstable vector meson state ρ at around 750 Mev, which decays into 2π or 4π. For simplicity, let us ignore the rather large width ($\Gamma_\rho \sim 100$ Mev) of the ρ state and treat it as stable. The mathematical complications resulting from the instability are not severe and have been discussed elsewhere.[9,10]

In place of (2.4), then, we have the definition

$$\langle 0|\mathfrak{J}_\alpha|\rho\rangle = m_\rho^2 (2\gamma_\rho)^{-1}\phi_\alpha, \qquad (2.10)$$

of the constant γ_ρ, where ϕ_α is the wave function of the ρ meson. In place of (2.1) or (2.3), we consider the matrix element between nucleon states of the isotopic spin current:

$$\langle N|\mathfrak{J}_\alpha|N\rangle = \bar{u}_f i\gamma_\alpha(\tau/2)u_i F_1^V(s) + \text{magnetic term}, \quad (2.11)$$

where $F_1^V(s)$ is the familiar isovector form factor of the electric charge of the nucleon, since the electromagnetic current has the form

$$j_\alpha = \mathfrak{J}_{3\alpha} + \text{isoscalar term}. \qquad (2.12)$$

If we continue to ignore the width of ρ, we get a dispersion relation like (2.6) with a pole term at m_ρ^2:

$$F_1^V(s) = (\gamma_{NN\rho}/\gamma_\rho) m_\rho^2 (m_\rho^2 - s)^{-1}$$
$$+ \int \sigma_1^V(M^2) dM^2 \, M^2 (M^2 - s - i\epsilon)^{-1}. \quad (2.13)$$

Here $\gamma_{NN\rho}$ is the coupling constant of ρ to $\bar{u}_f i\tau\gamma_\alpha u_i$, just as $g_{NN\pi}$ is the coupling constant of π to $\bar{u}_f i\tau\gamma_5 u_i$. In this case, we have used an unsubtracted dispersion relation just for convenience.

Since the current is conserved, there is no renormalization and we have

$$F_1^V(0) = 1, \qquad (2.14)$$

giving, in place of (2.7), the sum rule

$$1 = \gamma_{\rho NN}/\gamma_\rho + \int \sigma_1^V(M^2) dM^2. \qquad (2.15)$$

If the dispersion relation is dominated at low s by the ρ term, then we obtain the analog of the Goldberger-Treiman formula:

$$1 \approx \gamma_{\rho NN}/\gamma_\rho. \qquad (2.16)$$

---

[9] G. F. Chew, University of California Radiation Laboratory Report No. UCRL-9289, 1960 (unpublished).
[10] M. Gell-Mann and F. Zachariasen, Phys. Rev. 124, 953 (1961).

Now the same reasoning may be applied to the isovector electric form factor of another particle, for example the pion:

$$\langle \pi | \mathfrak{J}_\alpha | \pi \rangle = [i\phi_f^* \times \partial_\alpha \phi_i - i \partial_\alpha \phi_f^* \times \phi_i] F_\pi(s), \quad (2.17)$$

$$F_\pi(s) = (\gamma_{\rho\pi\pi}/\gamma_\rho) m_\rho^2 (m_\rho^2 - s)^{-1}$$
$$+ \int \sigma_\pi(M^2) dM^2 \, M^2 (M^2 - s - i\epsilon)^{-1}, \quad (2.18)$$

and

$$1 = \gamma_{\rho\pi\pi}/\gamma_\rho + \int \sigma_\pi(M^2) dM^2. \quad (2.19)$$

If this dispersion relation, too, is dominated by the ρ pole at low s, then we find

$$1 \approx \gamma_{\rho\pi\pi}/\gamma_\rho. \quad (2.20)$$

To the extent that the ρ pole gives most of the sum rule in each case, we have ρ coupled *universally* to the isotopic spins of nucleon, pion, etc., with coupling parameter $2\gamma_\rho$. Such universality was postulated by Sakurai,[11] within the framework of a special theory, in which ρ is treated as an elementary vector meson described by a Yang-Mills field. It can be seen that whether or not such a field description is correct, the *effective* universality ($\gamma_{\rho\pi\pi} \approx \gamma_{\rho NN} \approx \gamma_{\rho KK}$, etc.) is an approximate rule the validity of which depends on the domination of (2.15), (2.19), etc., by the ρ term.

The various coupling parameters $\gamma_{\rho\pi\pi}$, $\gamma_{\rho NN}$, etc., can be determined from the contribution of the ρ "pole" to various scattering processes, for example $\pi + N \to \pi + N$. But the factors $\gamma_{\rho\pi\pi}/\gamma_\rho$, $\gamma_{\rho NN}/\gamma_\rho$, etc., can also be measured, using electromagnetic interactions.[10]

An approximate determination of $\gamma_{\rho NN}/\gamma_\rho$ was made by Hofstadter and Herman[12] as follows The masses M^2 in the integral in Eq. (2.13) are taken to be effectively vary large, so that (2.13) becomes approximately

$$F_1{}^V(s) \approx (\gamma_{NN\rho}/\gamma_\rho) m_\rho^2 (m_\rho^2 - s)^{-1}$$
$$+ 1 - (\gamma_{\rho NN}/\gamma_\rho). \quad (2.21)$$

Fitting the experimental data on $F_1{}^V(s)$ with such a formula and using $m_\rho \approx 750$ Mev, we obtain $\gamma_{\rho NN}/\gamma_\rho \approx 1.4$. (Hofstadter and Herman, with a smaller value of m_ρ, found 1.2.)

III. EQUAL-TIME COMMUTATION RELATIONS

The dispersion relations for the matrix elements of weak or electromagnetic currents are linear and homogeneous. For example, Eq. (2.6) may be thought of as an expression for the matrix element of $\mathbf{P}_\alpha$ between the vacuum and a nucleon-antinucleon pair state. On the right-hand side, the pole term contains the product of the matrix element of $\mathbf{P}_\alpha$ between the vacuum and a one-pion state multiplied by the transition amplitude for the transition from π to $N\bar{N}$ by means of the strong interactions. The weight function $\sigma_{ax}(M^2)$ is just the sum of such products over many intermediate states (such as 3π, 5π, etc.) with total mass M.

Now such linear, homogeneous equations may determine the dependence of the current matrix elements on variables such as s, but they cannot fix the scale of these matrix elements; constants like $-G_A/G$ cannot be calculated without further information. A field theory of the strong interactions, with explicit expressions for the currents, somehow contains more than these dispersion relations. In what follows, we shall extract some of this additional information in the form of equal-time commutation relations between components of the currents. Since these are nonlinear relations, they can help to fix the scale of each matrix element. Moreover, these relations may be the same for the lepton system and for the baryon-meson system, so that universality of strength of the weak interactions, for example, becomes meaningful.[13]

Let us begin our discussion of equal-time commutation relations with a familiar case—that of the isotopic spin $\mathbf{I}$. Its components I_i obey the well-known commutation relations

$$[I_i, I_j] = i e_{ijk} I_k. \quad (3.1)$$

In terms of the components $\mathfrak{J}_{i\alpha}$ of the isotopic spin current, we have

$$I_i = -i \int \mathfrak{J}_{i4} d^3x, \quad (3.2)$$

and the conservation law

$$\partial_\alpha \mathfrak{J}_{i\alpha} = 0 \quad (3.3)$$

tells us that

$$\dot{I}_i = \int \partial_\alpha \mathfrak{J}_{i\alpha} d^3x = 0, \quad (3.4)$$

at all times.

Now the commutator of $\mathfrak{J}_{i4}(\mathbf{x},t)$ and $\mathfrak{J}_{j4}(\mathbf{x}',t)$ must vanish for $\mathbf{x} \neq \mathbf{x}'$, in accorance with microcausality. (Note we have taken the times equal.) If the commutator is not more singular than a delta function, then (3.1) and (3.2) give us the relation

$$[\mathfrak{J}_{i4}(\mathbf{x},t), \mathfrak{J}_{j4}(\mathbf{x}',t)] = -i e_{ijk} \mathfrak{J}_{k4}(\mathbf{x},t) \delta(\mathbf{x}-\mathbf{x}'), \quad (3.5)$$

which can also be obtained in any simple field theory by explicit commutation.[14]

In discussing the various parts of the weak current J_α, we shall have to deal with currents like $\mathbf{P}_\alpha$ that are not

---

[11] J J Sakurai, Ann. Phys. 11, 1 (1960).
[12] R. Hofstadter and R. Herman, Phys. Rev. Letters 6, 293 (1961). See also S. Bergia, A. Stanghellini, S. Fubini, and C. Villi, Phys. Rev. Letters 6, 367 (1961).

[13] M. Gell-Mann, *Proceedings of the 1960 Annual International Conference on High-Energy Physics at Rochester* (Interscience Publishers, Inc., New York, 1960).
[14] In some cases explicit commutation may be ambiguous and misleading. For example, a superficial consideration of $[j_i(\mathbf{x},t), j_4(\mathbf{x}',t)]$ for $i = 1, 2, 3$ may lead to the conclusion that the expression vanishes. Yet the vacuum expectation value of the commutator can be shown to be a nonzero quantity times $\partial_i \delta(\mathbf{x}-\mathbf{x}')$, and that result is confirmed by more careful calculation. See J. Schwinger, Phys. Rev. Letters 3, 296 (1959).

conserved.[16] Here, too, we may define a quantity analogous to **I**:

$$D_i = -i \int P_{i4} d^3x, \qquad (3.6)$$

but D_i is *not* independent of time:

$$\dot{D}_i = \int \partial_\alpha P_{i\alpha} d^3x \neq 0. \qquad (3.7)$$

For the moment, let us restrict our attention to the currents $\mathfrak{J}_\alpha$ and $\mathbf{P}_\alpha$ and the operators **I** and $\mathbf{D}(t)$. Since **D** is an isovector, we have the relations

$$[I_i, D_j] = [D_i, I_j] = i e_{ijk} D_k, \qquad (3.8)$$

but what is the commutator of two components of **D**? Since $\mathbf{P}_\alpha$ is a physical quantity, so is **D** and the question is one with direct physical meaning. We shall give both a general and a specific answer.

In general, we may take the commutators of D's (divided by i), the components of **I** and **D**, the commutators of all of these with one another (divided by i), etc., until we obtain a system of Hermitian operators that is closed under commutation. Any of these operators can be written as a linear combination of N linearly independent Hermitian operators $R_i(t)$, where N might be infinite, and where the commutator of any two R_i is a linear combination of the R_i:

$$[R_i(t), R_j(t)] = i c_{ijk} R_k(t), \qquad (3.9)$$

with c_{ijk} real. Such a system is called an algebra by the mathematicians. If we consider the set of infinitesimal unitary operators $1 + i\epsilon R_i(t)$ and all possible products of these, we obtain an N-parameter continuous group of unitary transformations. We can refer to (3.9) as the algebra of the group. It is a physically meaningful statement to specify what group or what algebra is generated in this way by the currents $\mathfrak{J}_\alpha$ and $\mathbf{P}_\alpha$. Since a commutation relation like (3.9) is left invariant by a unitary transformation such as $\exp(-it\int H d^3x)$, the numbers c_{ijk} are independent of time.

A second mathematical statement is also in order, i.e., the specification of the transformation properties of the Hamiltonian density $H(\mathbf{x},t)$ under the group or the algebra. Those R_i for which $[R_i(t), H(\mathbf{x},t)] = 0$ are independent of time, but some of them, like D_i, do not commute with H. If all of the R_i commuted with H, then H would belong to the trivial one-dimensional representation of the group. In fact, H behaves in a more complicated way. By commuting all of the $R_i(t)$ with $H(\mathbf{x},t)$, we obtain a linear set of operators, containing H, that form a representation of the group; it may be broken up into the direct sum of irreducible representations. We want to know, then, what group is generated by **I** and **D** and to what irreducible representations of this group H belongs. Suggested are specific answers to both questions.

Let us look at the vector and axial vector weak currents for the leptons. For the time being, we shall consider only ν and e, ignoring the muon. (In the same way, we shall, in this section, ignore strange particles, and consider only baryons and mesons with $S=0$.) The vector weak current $i\bar{\nu}\gamma_\alpha e$ and the axial current $i\bar{\nu}\gamma_\alpha\gamma_5 e$ can be regarded formally as components of two "isotopic vector" currents for the leptons:

$$\mathfrak{J}_\alpha{}^{(l)} = i\bar{\xi}\tau\gamma_\alpha \xi/2, \quad \mathbf{P}_\alpha{}^{(l)} = i\bar{\xi}\tau\gamma_\alpha\gamma_5 \xi/2, \qquad (3.10)$$

where ξ stands for (ν, e). We can also form the mathematical analogs of **I** and **D**:

$$\mathbf{I}^{(l)} = -i \int \mathfrak{J}_\alpha{}^{(l)} d^3x, \quad \mathbf{D}^{(l)} = -i \int \mathbf{P}_\alpha{}^{(l)} d^3x. \qquad (3.11)$$

Now in this leptonic case we can easily compute the commutation rules of $\mathbf{I}^{(l)}$ and $\mathbf{D}^{(l)}$:

$$[I_i{}^{(l)}, I_j{}^{(l)}] = i e_{ijk} I_k{}^{(l)},$$
$$[I_i{}^{(l)}, D_j{}^{(l)}] = [D_i{}^{(l)}, I_j{}^{(l)}] = i e_{ijk} D_k{}^{(l)}, \qquad (3.12)$$
$$[D_i{}^{(l)}, D_j{}^{(l)}] = i e_{ijk} I_k{}^{(l)}.$$

Another way to phrase these commutation rules is to put

$$\mathbf{I}^{(l)} = \mathbf{I}_+{}^{(l)} + \mathbf{L}^{-(l)},$$
$$\mathbf{D}^{(l)} = \mathbf{I}_+{}^{(l)} - \mathbf{L}^{-(l)}, \qquad (3.13)$$

and to notice that $\mathbf{I}_+{}^{(l)}$ and $\mathbf{L}^{-(l)}$ are two commuting angular momenta [essentially $\tau(1+\gamma_5)/4$ and $\tau(1-\gamma_5)/4$]. The weak current $i\bar{\nu}\gamma_\alpha(1+\gamma_5)e$ is just a component of the current of $\mathbf{I}_+{}^{(l)}$.

We now suggest that the algebraic structure of **I** and **D** is exactly the same in the case of baryons and mesons. To (3.1) and (3.8), we add the rule[16,17]

$$[D_i, D_j] = i e_{ijk} I_k, \qquad (3.14)$$

which closes the system and makes $\mathbf{I}^+ \equiv (\mathbf{I}+\mathbf{D})/2$ and $\mathbf{I}^- \equiv (\mathbf{I}-\mathbf{D})/2$ two commuting angular momenta. Again, we make the weak current a component of the current of $\mathbf{I}^+$. Evidently the statement that $(\mathbf{I}+\mathbf{D})/2$ is an angular momentum and not some factor times an angular momentum, fixes the scale of the weak current. It makes universality of strength between baryons and leptons meaningful, and it specifies, together with the dispersion relations, the value of such constants as $-G_A/G$.

The simplest way to realize the algebraic structure under discussion in a field-theory model of baryons and mesons is to construct the currents $\mathfrak{J}_\alpha$ and $\mathbf{P}_\alpha$ out of p and n fields just as $\mathfrak{J}_\alpha{}^{(l)}$ and $\mathbf{P}_\alpha{}^{(l)}$ are made out of ν and e fields:

$$\mathfrak{J}_\alpha = i\bar{N}\tau\gamma_\alpha N/2, \quad \mathbf{P}_\alpha = i\bar{N}\tau\gamma_\alpha\gamma_5 N/2, \qquad (3.15)$$

---

[16] We assume that the vector weak current with $\Delta S = 0$ is just a component of the isotopic spin current $\mathfrak{J}_\alpha$ and thus conserved.

[16] F. Gursey, Nuovo cimento **16**, 230 (1960).
[17] M. Gell-Mann and M. Lévy, Nuovo cimento **16**, 705 (1960).

where N means (p,n). We then obtain not only the commutation rules (3.1), (3.8), and (3.14), but the stronger rule (3.5) and its analogs:

$$[\mathfrak{J}_{i4}(\mathbf{x},t), P_{j4}(\mathbf{x}',t)] = -ie_{ijk}P_{k4}(\mathbf{x},t)\delta(\mathbf{x}-\mathbf{x}'),$$
$$[P_{i4}(\mathbf{x},t), P_{j4}(\mathbf{x}',t)] = -ie_{ijk}\mathfrak{J}_{k4}(\mathbf{x},t)\delta(\mathbf{x}-\mathbf{x}'). \quad (3.16)$$

Next we want to use a field-theory model to suggest an answer to the second question—how H behaves under the group or, what is the same thing, under the algebra consisting of $\mathbf{I}$ and $\mathbf{D}$ or of $\mathbf{I}^+$ and $\mathbf{I}^-$. Since $\mathbf{I}^+$ and $\mathbf{I}^-$ are two commuting angular momenta, any irreducible representation of the algebra is specified by a pair of total angular momentum quantum numbers: i_+ for $\mathbf{I}^+$ and i_- for $\mathbf{I}^-$. The total isotopic spin quantum number I is associated with $\mathbf{I}^+ + \mathbf{I}^- = \mathbf{I}$.

Now we want the vector weak current $\mathfrak{J}_\alpha$ to be the isotopic spin current and to be conserved. Thus H must commute with $\mathbf{I}$; it transforms as an isoscalar, with $I=0$. In order to couple to zero, i_+ and i_- must be equal. So H can consist of terms with $(i_+, i_-) = (0,0)$, $(\tfrac{1}{2},\tfrac{1}{2})$, $(1,1)$, $(\tfrac{3}{2},\tfrac{3}{2})$, etc. Which of these are in fact present?

The simplest model in which the total isotopic current is given by just (3.15) is the Fermi-Yang[18] model, in which the pion is a composite of nucleon and antinucleon. To write an explicit Lagrangian, it must be decided what form the binding interaction takes. Since a direct four-fermion coupling leads to unpleasant singularities, we whall use a massive neutral vector meson field B^0 coupled to the nucleon current, as proposed by Teller[18] and Sakurai[11]; the exchange of a B^0 gives attraction between nucleon and antinucleon, permitting binding, and it also gives repulsion between nucleons, contributing to the "hard core." The model Lagrangian is then[19]

$$L = -\bar{N}\gamma_\alpha \partial_\alpha N - (\partial_\alpha B_\beta - \partial_\beta B_\alpha)^2/4$$
$$-\mu_0^2 B_\alpha B_\alpha/2 - ih_0 B_\alpha \bar{N}\gamma_\alpha N - m_0 \bar{N}N. \quad (3.17)$$

If the mass term for the nucleon were absent, then both $\mathfrak{J}_\alpha$ and $\mathbf{P}_\alpha$ would be conserved; $\mathbf{I}$ and $\mathbf{D}$ would both commute with L and with H. Thus,

$$H = H(0,0) - u_0, \quad (3.18)$$

where $H(0,0)$ transforms according to $(i_+, i_-) = (0,0)$ and the noninvariant term u_0 is just $-m_0\bar{N}N$. To what representation does it belong?

It is easy to see that the field B^0 belongs to $(0,0)$, while $N_L \equiv (1+\gamma_5)N/2$ belongs to $(\tfrac{1}{2},0)$ and N_R

---

[18] E. Fermi and C. N. Yang, Phys. Rev. 76, 1739 (1949); E. Teller, *Proceedings of the Sixth Annual Rochester Conference on High-Energy Nuclear Physics, 1956* (Interscience Publishers, Inc., New York, 1956).

[19] Conceivably a massive B^0 meson can be described by (3.17) even with $\mu_0 = 0$. [J. Schwinger, lectures at Stanford University, summer, 1961 (unpublished)]. In that case the noninvariant term in (3.17) is just equal to $\theta_{\alpha\alpha}$ and the traceless part of $\theta_{\alpha\beta}$ commutes with the group elements at equal times. In any case, whether μ_0 is zero or not, the off-diagonal terms in $\theta_{\alpha\beta}$ commute with the group.

$\equiv (1-\gamma_5)N/2$ belongs to $(0,\tfrac{1}{2})$. One can thus verify that all terms of (3.17) except the last belong to $(0,0)$, since $\bar{N}\gamma_\alpha N$ or $\bar{N}\gamma_\alpha \partial_\alpha N$ couples $\bar{N}_L$ to N_L and $\bar{N}_R$ to N_R. But the Dirac matrix β, unlike $\beta\gamma_\alpha$, anticommutes with γ_5, so that the last term $-m_0\bar{N}N$ couples $\bar{N}_L$ to N_R and $\bar{N}_R$ to N_L. Thus u_0 belongs to $(\tfrac{1}{2},\tfrac{1}{2})$. We have $H = H(0,0) + H(\tfrac{1}{2},\tfrac{1}{2})$.

There are four components to the representation $(\tfrac{1}{2},\tfrac{1}{2})$ to which $u_0 = -H(\tfrac{1}{2},\tfrac{1}{2})$ belongs. By commuting $\mathbf{D}$ with u_0, we generate the other three easily and see that they are proportional to $-i\bar{N}\gamma_5 \boldsymbol{\tau} N$. In fact $\mathbf{D}$ acts like $\boldsymbol{\tau}\gamma_5/2$, $\mathbf{I}$ like $\boldsymbol{\tau}/2$, u_0 like β, and the other three components like $-i\beta\gamma_5\boldsymbol{\tau}$. Denoting the three new components by v_i, we have

$$[I_i, u_0] = 0, \quad [D_i, u_0] = -iv_i,$$
$$[I_i, v_j] = ie_{ijk}v_k, \quad [D_i, v_j] = i\delta_{ij}u_0. \quad (3.19)$$

In the model, there are the even stronger relations for the densities

$$[\mathfrak{J}_{i4}(\mathbf{x},t), u_0(\mathbf{x}',t)] = 0, \quad [P_{i4}(\mathbf{x},t), u_0(\mathbf{x}',t)]$$
$$= -iv_i(\mathbf{x},t)\delta(\mathbf{x}-\mathbf{x}'), \text{ etc.} \quad (3.20)$$

The noninvariant term u_0 is what prevents the axial vector current from being conserved. Thus one can express the divergence $\partial_\alpha P_\alpha$ of the current in terms of the commutator of $\mathbf{D}$ with u_0. The conditions for this relation to hold are treated in the appendix and are applicable to all models we discuss. We find simply

$$\partial_\alpha P_\alpha = -i[\mathbf{D}, H] = i[\mathbf{D}, u_0] = \mathbf{v}, \quad (3.21)$$

and, of course,

$$\partial_\alpha \mathfrak{J}_\alpha = -i[\mathbf{I}, H] = 0. \quad (3.22)$$

It is precisely the operator $\mathbf{v}$, then, that we used in a dispersion relation in order to obtain the Goldberger-Treiman relation in Sec. II. Acting on the vacuum, it leads mostly to the one-pion state, so that the pion is effectively coupled universally to the divergence of the axial vector current. Thus $\mathbf{v}$ is a sort of effective pion field operator for the Fermi-Yang theory, which has no explicit pion field.

If we insist on a model in which there is a field variable $\pi(x,t)$ then we must complicate the discussion. The total isotopic spin current is no longer given by just (3.15); there is a pion isotopic current term as well. In order to preserve the same algebraic structure of $\mathbf{I}$ and $\mathbf{D}$, one must then modify $\mathbf{P}_\alpha$ as well. Such a theory was described by Gell-Mann and Lévy,[17] who called it the "σ-model".[20] Along with the field $\boldsymbol{\pi}$, we must introduce a scalar, isoscalar field σ' in such a way that $\boldsymbol{\pi}$, σ' transform under the group like $\mathbf{v}$, u_0. Then, just as $\mathfrak{J}_\alpha$ has an additional term quadratic in $\boldsymbol{\pi}$, $\mathbf{P}_\alpha$ requires an additional term bilinear in $\boldsymbol{\pi}$ and σ'.

As we shall see in the next section, the introduction of

---

[20] In the σ model, explicit commutation of u_0 and $\mathbf{v}$ at equal times gives zero, while in the Fermi-Yang model this is not so; if we take these results seriously, they give us definite physical distinctions among models.

strange particles makes the group much larger. The term u_0 is then a member of a much larger representation, with eighteen components. Thus if a pion field is introduced, fifteen more components are needed as well. Such a theory is too complicated to be attractive; we shall therefore ignore it and concentrate on the simplest generalization of the Fermi-Yang model to strange particles, namely the symmetrical Sakata model.

IV. SYMMETRICAL SAKATA MODEL AND UNITARY SYMMETRY

In the previous section, we proceeded inductively. We showed that starting from physical currents like $\mathfrak{J}_\alpha$ and $\mathbf{P}_\alpha$ we may construct a group and its algebra and that it is physically meaningful to specify the group and also the transformation properties of H under the group. We chose the algebraic structure by analogy with the case of leptons and we saw that the simplest field theory model embodying the structure is just the Fermi-Yang model, in which p and n fields are treated just like the ν and e fields for the leptons, except that they are given a mass and a strong "gluon" coupling. The transformation properties of H were taken from the model; H consists, then, of an invariant part $H_{0,0}$ plus a term $(-u_0)$, where u_0 and a pseudoscalar isovector quantity $\mathbf{v}$ belong to the representation $(\frac{1}{2},\frac{1}{2})$ of the group. We then have the commutation rules (3.1), (3.8), (3.14), and (3.19). Microcausality with the assumption of commutators that are not too singular, or else direct inspection of the model, gives the stronger commutation rules (3.5), (3.16), and (3.20) for the densities. The model also gives specific equal-time commutation rules for u_0 and $\mathbf{v}$, which we did not list. All of these properties can be abstracted from the model and considered on their own merits as proposed relations among the currents and the Hamiltonian density.

Now, to argue deductively, we want to include the strange particles and all parts of the weak current J_α and the electromagnetic current j_α. We generalize the Fermi-Yang description to obtain the symmetrical Sakata model and abstract from it as many physically meaningful relations as possible.

It has long been recognized that the qualitative properties of baryons and mesons could be understood in terms of the Sakata model,[21] in which all strongly interacting particles are made out of N, Λ, $\bar{N}$, and $\bar{\Lambda}$ (or at least out of basic fields with the same quantum numbers as these particles).

We write the Lagrangian density for the Sakata model as a generalization of (3.17):

$$L = -\bar{p}\gamma_\alpha\partial_\alpha p - \bar{n}\gamma_\alpha\partial_\alpha n - \bar{\Lambda}\gamma_\alpha\partial_\alpha\Lambda - \tfrac{1}{4}(\partial_\alpha B_\beta - \partial_\beta B_\alpha)^2 \\ - \tfrac{1}{2}\mu_0^2 B_\alpha B_\alpha - ih_0(\bar{p}\gamma_\alpha p + \bar{n}\gamma_\alpha n + \bar{\Lambda}\gamma_\alpha\Lambda)B_\alpha \\ - m_{0N}(\bar{n}n + \bar{p}p) - m_{0\Lambda}\bar{\Lambda}\Lambda. \quad (4.1)$$

According to this picture, the baryons present a striking parallel with the leptons,[22] for which we write the Lagrangian density

$$L_l = -\bar{\nu}\gamma_\alpha\nu - \bar{e}\gamma_\alpha\partial_\alpha e - \bar{\mu}\gamma_\alpha\partial_\alpha\mu - O\cdot(\bar{\nu}\nu + \bar{e}e) - m_\mu\bar{\mu}\mu, \quad (4.2)$$

if we turn off the electromagnetic and weak couplings, along with the ν-e mass difference. Here it is assumed there is only one kind of neutrino.

The only real difference between baryons and leptons in (4.1) and (4.2), respectively, is that the baryons are coupled, through the baryon current, to the field B. It is tempting to suppose that the weak current of the strongly interacting particles is just the expression.

$$i\bar{p}\gamma_\alpha(1+\gamma_5)n + i\bar{p}\gamma_\alpha(1+\gamma_5)\Lambda, \quad (4.3)$$

analogous to Eq. (1.2) for the leptonic weak current $J_\alpha^{(l)}$. Now (4.3) is certainly a reasonable expression, qualitatively, for weak currents of baryons and mesons. As Okun has emphasized,[23] the following properties of the weak interactions, often introduced as postulates, are derivable from (1.1), (1.2), (4.1), and (4.3):

(a) The conserved vector current.[24] In the model under discussion, as in that of Fermi and Yang, $i\bar{p}\gamma_\alpha n$ is a component of the total isotopic spin current.

(b) The rules $|\Delta S| = 1$, $\Delta S/\Delta Q = +1$, and $|\Delta \mathbf{I}| = \tfrac{1}{2}$ for the leptonic decays of strange particles.[25]

(c) The invariance under GP of the $\Delta S = 0$ weak current.[26]

(d) The rules $|\Delta S| = 1$, $|\Delta \mathbf{I}| = \tfrac{1}{2}$ or $\tfrac{3}{2}$ in the nonleptonic decays of strange particles; along with $|\Delta S| = 1$, we have the absence of a large K_1^0-K_2^0 mass difference.

The quantitative facts that the effective coupling constants for $|\Delta S| = 1$ leptonic decays are smaller than those for $|\Delta S| = 0$ leptonic decays and that in nonleptonic decays of strange particles the $|\Delta \mathbf{I}| = \tfrac{1}{2}$ amplitude greatly predominates over the $|\Delta \mathbf{I}| = \tfrac{3}{2}$ amplitude are not explained in any fundamental way.[27]

---

[21] S. Sakata, Progr. Theoret. Phys. (Kyoto) **16**, 686 (1956).

[22] A. Gamba, R. E. Marshak, and S. Okubo, Proc. Natl. Acad. Sci. U. S. **45**, 881 (1959).

[23] L. Okun, Ann. Rev. Nuclear Sci. **9**, 61 (1959).

[24] R. P. Feynman and M. Gell-Mann, Phys. Rev. **109**, 193 (1958). See also S. S. Gershtein and J. B. Zeldovich, Soviet Phys.—JETP **2**, 576 (1957).

[25] M. Gell-Mann, *Proceedings of the Sixth Annual Rochester Conference on High-Energy Nuclear Physics, 1956* (Interscience Publishers, Inc., New York, 1956). These rules were in fact suggested on the basis of the idea that N and Λ are fundamental. Should the rules prove too restrictive (for example should $\Delta S/\Delta Q = +1$ be violated), then we would try a larger group; in the language of the field-theoretic model, we would assume more fundamental fields. For a discussion of possible larger groups, see M. Gell-Mann and S. Glashow, Ann. Phys. **15**, 437 (1961) and S. Coleman and S. Glashow (to be published).

[26] S. Weinberg, Phys. Rev. **112**, 1375 (1958).

[27] A possible dynamical explanation of the predominance of $|\Delta \mathbf{I}| = \tfrac{1}{2}$ is being investigated by Nishijima (private communication). For example, consider the decay $\Lambda \rightarrow N + \pi$. A dispersion relation without subtractions is written for the matrix element of $J_\alpha^\dagger J_\alpha$ between the vacuum and a state containing $N + \bar{\Lambda} + \pi$. The parity-violating part leads to intermediate pseudoscalar states with $S = +1$ and with $|\Delta \mathbf{I}| = \tfrac{1}{2}$ or $\tfrac{3}{2}$. In the case of $|\Delta \mathbf{I}| = \tfrac{1}{2}$, there is an intermediate K particle, which may give a large contribution, swamping the term with $|\Delta \mathbf{I}| = \tfrac{3}{2}$, which has no one-meson state. For the same argument to apply to the parity-conserving part, we need the K' meson of Table III.

TABLE I. A set of matrices λ_i.

$$\lambda_1 = \begin{pmatrix} 0 & 1 & 0 \\ 1 & 0 & 0 \\ 0 & 0 & 0 \end{pmatrix} \quad \lambda_2 = \begin{pmatrix} 0 & -i & 0 \\ i & 0 & 0 \\ 0 & 0 & 0 \end{pmatrix} \quad \lambda_3 = \begin{pmatrix} 1 & 0 & 0 \\ 0 & -1 & 0 \\ 0 & 0 & 0 \end{pmatrix}$$

$$\lambda_4 = \begin{pmatrix} 0 & 0 & 1 \\ 0 & 0 & 0 \\ 1 & 0 & 0 \end{pmatrix} \quad \lambda_5 = \begin{pmatrix} 0 & 0 & -i \\ 0 & 0 & 0 \\ i & 0 & 0 \end{pmatrix} \quad \lambda_6 = \begin{pmatrix} 0 & 0 & 0 \\ 0 & 0 & 1 \\ 0 & 1 & 0 \end{pmatrix}$$

$$\lambda_7 = \begin{pmatrix} 0 & 0 & 0 \\ 0 & 0 & -i \\ 0 & i & 0 \end{pmatrix} \quad \lambda_8 = \begin{pmatrix} 1/\sqrt{3} & 0 & 0 \\ 0 & 1/\sqrt{3} & 0 \\ 0 & 0 & -2/\sqrt{3} \end{pmatrix}$$

The electromagnetic properties of baryons and leptons are not exactly parallel in the Sakata model. The electric current (divided by e), which are denoted by j_α, is given by

$$i\bar{p}\gamma_\alpha p \tag{4.4}$$

for the baryons and mesons and by

$$-i(\bar{e}\gamma_\alpha e + \bar{\mu}\gamma_\alpha \mu) \tag{4.5}$$

for the leptons.

Now, we return to the Lagrangian (4.1) and separate it into three parts:

$$L = \bar{L} + L' + L'', \tag{4.6}$$

where $\bar{L}$ stands for everything except the baryon mass terms, while L' and L'' are given by the expressions

$$L' = (2m_{0N} + m_{0\Lambda})(\bar{N}N + \bar{\Lambda}\Lambda)/3,$$
$$L'' = (m_{0N} - m_{0\Lambda})(\bar{N}N - 2\bar{\Lambda}\Lambda)/3. \tag{4.7}$$

If we now consider the Lagrangian with the mass-splitting term L'' omitted, we have a theory that is completely symmetrical in p, n, and Λ. We may perform any unitary linear transformation (with constant coefficients) on these three fields and leave $\bar{L} + L'$ invariant. Thus in the absence of the mass-splitting term L'' the theory is invariant under the three-dimensional unitary group $U(3)$; we shall refer to this situation as "unitary symmetry."

If we now turn on the mass-splitting, the symmetry is reduced. The only allowed unitary transformations are those involving n and p alone or Λ alone. The group becomes $U(2) \times U(1)$, which corresponds, as we shall see, to the conservation if isotopic spin, strangeness, and baryon number.

For simplicity, let us return briefly to the simpler case in which there is no Λ. The symmetry group is then just $U(2)$, the set of unitary transformations on n and p. We can factor each unitary transformation uniquely into one which multiplies both fields by the same phase factor and one (with determinant unity) which leaves invariant the product of the phase factors of p and n. Invariance under the first kind of transformation corresponds to conservation of nucleons n and p; it may be considered separately from invariance under the class of transformations of the second kind [called by mathematicians the unitary unimodular group $SU(2)$ in two dimensions]. In mathematical language, we can factor $U(2)$ into $U(1) \times SU(2)$.

Each transformation of the first kind can be written as a matrix $1 \exp i\phi$, where 1 is the unit 2×2 matrix. The infinitesimal transformation is $1 + i 1 \delta\phi$, and so the unit matrix is the infinitesimal generator of these transformations. Those of the second kind are generated in the same way by the three independent traceless 2×2 matrices, which may be taken to be the Pauli isotopic spin matrices τ_1, τ_2, and τ_3. We thus have

$$N \to (1 + i \sum_{k=1}^{3} \delta\theta_k \tau_k/2)N, \tag{4.8}$$

as the general infinitesimal transformation of the second kind. Symmetry under all the transformations of the second kind is the same as symmetry under isotopic spin rotations. The whole formalism of isotopic spin theory can then be constructed by considering the transformation properties of the doublet or spinor (p,n) and of more complicated objects that transform like combinations of two or more such nucleons (or antinucleons).

The Pauli matrices τ_k are Hermitian and obey the rules

$$\begin{aligned} \mathrm{Tr}\,\tau_i\tau_j &= 2\delta_{ij}, \\ [\tau_i, \tau_j] &= 2ie_{ijk}\tau_k, \\ \{\tau_i, \tau_j\} &= 2\delta_{ij}1. \end{aligned} \tag{4.9}$$

The invariance under the group $SU(2)$ of isotopic spin rotations corresponds to conservation of the isotopic spin current

$$\mathfrak{J}_\alpha = i\bar{N}\tau\gamma_\alpha N/2,$$

while the invariance under transformations of the first kind corresponds to conservation of the nucleon current $i\bar{N}\gamma_\alpha N/2 = n_\alpha$.

Defining the total isotopic spin $\mathbf{I}$ as in (3.2), we obtain for I_i the commutation rules (3.1), which are the same as those for $\tau_i/2$. Likewise the nucleon number is defined as $-i \int n_4 d^3x$ and commutes with $\mathbf{I}$.

We now generalize the idea of isotopic spin by including the third field Λ. Again we factor the unitary transformations on baryons into those which are generated by the 3×3 unit matrix 1 (and which correspond to baryon conservation) and those which are generated by the eight independent traceless 3×3 matrices [and which form the unitary unimodular group $SU(3)$ in three dimensions]. We may construct a typical set of eight such matrices by analogy with the 2×2 matrices of Pauli. We call then $\lambda_1 \cdots \lambda_8$ and list them in Table I. They are Hermitian and have the properties

$$\begin{aligned} \mathrm{Tr}\,\lambda_i\lambda_j &= 2\delta_{ij}, \\ [\lambda_i, \lambda_j] &= 2if_{ijk}\lambda_k, \\ \{\lambda_i, \lambda_j\} &= 2d_{ijk}\lambda_k + \tfrac{4}{3}\delta_{ij}1, \end{aligned} \tag{4.10}$$

where f_{ijk} is real and totally antisymmetric like the

TABLE II. Nonzero elements of f_{ijk} and d_{ijk}. The f_{ijk} are odd under permutations of any two indices while the d_{ijk} are even.

ijk	f_{ijk}	ijk	d_{ijk}
123	1	118	$1/\sqrt{3}$
147	1/2	146	1/2
156	−1/2	157	1/2
246	1/2	228	$1/\sqrt{3}$
257	1/2	247	−1/2
345	1/2	256	1/2
367	−1/2	338	$1/\sqrt{3}$
458	$\sqrt{3}/2$	344	1/2
678	$\sqrt{3}/2$	355	1/2
...	...	366	−1/2
...	...	377	−1/2
...	...	448	$-1/(2\sqrt{3})$
...	...	558	$-1/(2\sqrt{3})$
...	...	668	$-1/(2\sqrt{3})$
...	...	778	$-1/(2\sqrt{3})$
...	...	888	$-1/\sqrt{3}$

Kronecker symbol e_{ijk} of Eq. (4.9), while d_{ijk} is real and totally symmetric. These properties follow from the equations

$$\mathrm{Tr}\lambda_k[\lambda_i,\lambda_j] = 4if_{ijk},$$
$$\mathrm{Tr}\lambda_k\{\lambda_i,\lambda_j\} = 4d_{ijk}, \quad (4.11)$$

derived from (4.10).

The nonzero elements of f_{ijk} and d_{ijk} are given in Table II for our choice of λ_i. Even and odd permutations of the listed indices correspond to multiplication of f_{ijk} by ± 1, respectively, and of d_{ijk} by $+1$.

The general infinitesimal transformation of the second kind on the three basic baryons b is, of course,

$$b \rightarrow (1 + i\sum_{i=1}^{8} \delta\theta_i\lambda_i/2)b, \quad (4.12)$$

by analogy with (4.8). Together with conservation of baryons, invariance under these transformations corresponds to complete "unitary symmetry" of the three baryons. We have factored $U(3)$ into $U(1)\times SU(3)$.

The invariance under transformations of the first kind gives us conservation of the baryon current

$$i\bar{b}\gamma_\alpha b = i\bar{n}\gamma_\alpha n + i\bar{p}\gamma_\alpha p + i\bar{\Lambda}\gamma_\alpha\Lambda, \quad (4.13)$$

while invariance under the second class of transformations would give us conservation of the eight-component "unitary spin" current

$$\mathcal{F}_{i\alpha} = i\bar{b}\lambda_i\gamma_\alpha b/2 \quad (i=1,\cdots,8). \quad (4.14)$$

Now in fact L'' is not zero and so not all the components of $\mathcal{F}_{i\alpha}$ are actually conserved. This does not prevent us from defining $\mathcal{F}_{i\alpha}$ as in (4.14), nor does it affect the commutation rules of the unitary spin density. The total unitary spin F_i is defined by the relation

$$F_i = -i\int \mathcal{F}_{i4}d^3x, \quad (4.15)$$

at any time and at equal times the commutation rules for F_i follow those for $\lambda_i/2$

$$[F_i, F_j] = if_{ijk}F_k. \quad (4.16)$$

The baryon number, of course, commutes with all components F_i.

It will be noticed that λ_1, λ_2, and λ_3 agree with τ_1, τ_2, and τ_3 for p and n and have no matrix elements for Λ. Thus the first three components of the unitary spin are just the components of the isotopic spin. The matrix λ_8 is diagonal in our representation and has one eigenvalue for the nucleon and another for the Λ. Thus F_8 is just a linear combination of strangeness and baryon number. It commutes with the isotopic spin.

The matrices λ_4, λ_5, λ_6, and λ_7 connect the nucleon and Λ. We see that the components F_4, F_5, F_6, and F_7 of the unitary spin change strangeness by one unit and isotopic spin by a half unit. When the mass-splitting term L'' is "turned on," it is these components that are no longer conserved, while the conservation of F_1, F_2, F_3, F_8, and baryon number remains valid.

V. VECTOR AND AXIAL VECTOR CURRENTS

We may unify the mathematical treatment of the baryon current and the unitary spin current if we define a ninth 3×3 matrix

$$\lambda_0 = (\tfrac{2}{3})^{\frac{1}{2}}\mathbf{1}, \quad (5.1)$$

so that the *nine* matrices λ_i obey the rules

$$[\lambda_i,\lambda_j] = 2if_{ijk}\lambda_k \quad (i=0,\cdots,8),$$
$$\{\lambda_i,\lambda_j\} = 2d_{ijk}\lambda_k \quad (i=0,\cdots,8), \quad (5.2)$$
$$\mathrm{Tr}\lambda_i\lambda_j = 2\delta_{ij} \quad (i=0,\cdots,8).$$

Here, f_{ijk} is defined as before, except that it vanishes when any index is zero; d_{ijk} is also defined as before, except that it has additional nonzero matrix elements equal to $(\tfrac{2}{3})^{\frac{1}{2}}$ whenever any index is zero and the other two indices are equal. The baryon current is now $(\tfrac{3}{2})^{\frac{1}{2}}\mathcal{F}_{0\alpha}$.

The definitions (4.15) and the equal-time commutation relations (4.16) now hold for $i=0,\cdots,8$. Moreover, there are the equal-time commutation relations

$$[\mathcal{F}_{i4}(\mathbf{x},t), \mathcal{F}_{j4}(\mathbf{x}',t)] = -if_{ijk}\mathcal{F}_{k4}(\mathbf{x},t)\delta(\mathbf{x}-\mathbf{x}') \quad (5.3)$$

for the densities.

The electric current j_α is then

$$j_\alpha = (\sqrt{2}\mathcal{F}_{0\alpha} + \mathcal{F}_{8\alpha} + \sqrt{3}\mathcal{F}_{3\alpha})/2\sqrt{3}, \quad (5.4)$$

while the vector weak current is

$$\mathcal{F}_{1\alpha} + i\mathcal{F}_{2\alpha} + \mathcal{F}_{4\alpha} + i\mathcal{F}_{5\alpha}. \quad (5.5)$$

We now wish to set up the same formalism for the axial vector currents. We recall that the presence of the symmetry-breaking term L'' did not prevent us from defining the $\mathcal{F}_{i\alpha}$ and obtaining the commutation rules (5.3) characteristic of the unitary symmetry group $U(3)$.

In the same way, we now remark that if both L''

and L' are "turned off," we have invariance under the infinitesimal unitary transformations

$$b \to (1+i\sum_{i=0}^{8} \delta\psi_i\gamma_5\lambda_i/2)b, \qquad (5.6)$$

as well as the infinitesimal transformations

$$b \to (1+i\sum_{i=0}^{8} \delta\theta_i\lambda_i/2)b \qquad (5.7)$$

we have used before.[28] Thus the axial vector currents

$$\mathfrak{F}_{i\alpha}{}^5 = i\bar{b}\lambda_i\gamma_5 b/2 \qquad (5.8)$$

would be conserved if both L' and L'' were absent. Even in the presence of these terms, we have the commutation rules

$$[\mathfrak{F}_{i4}{}^5(\mathbf{x},t), \mathfrak{F}_{j4}(\mathbf{x}',t)] = -if_{ijk}\mathfrak{F}_{k4}{}^5(\mathbf{x},t)\delta(\mathbf{x}-\mathbf{x}') \quad (5.9)$$

and

$$[\mathfrak{F}_{i4}{}^5(\mathbf{x},t), \mathfrak{F}_{j4}{}^5(\mathbf{x}',t)] = -if_{ijk}\mathfrak{F}_{k4}(\mathbf{x},t)\delta(\mathbf{x}-\mathbf{x}') \quad (5.10)$$

at equal times, We may use the definition

$$F_i{}^5(t) \equiv -i\int \mathfrak{F}_{i4}{}^5 d^3x, \qquad (5.11)$$

along with (4.15).

Just as we put $\mathbf{I} = \mathbf{I}_+ + \mathbf{L}$ and $\mathbf{D} = \mathbf{I}_+ - \mathbf{L}$ in the discussion following Eq. (3.16), so we now write

$$F_i(t) = F_i{}^+(t) + F_i{}^-(t),$$
$$F_i{}^5(t) = F_i{}^+(t) - F_i{}^-(t), \qquad (5.12)$$

and it is seen that $F_i{}^+$ and $F_i{}^-$ separately obey the commutation rules

$$[F_i{}^\pm, F_j{}^\pm] = if_{ijk}F_k{}^\pm, \qquad (5.13)$$

while they commute with each other:

$$[F_i{}^\pm, F_j{}^\mp] = 0. \qquad (5.14)$$

Thus we are now dealing with the group $U(3)$ taken twice: $U(3) \times U(3)$. Factoring each $U(3)$ into $U(1) \times SU(3)$, we have[29] $U(1) \times U(1) \times SU(3) \times SU(3)$. Thus we have defined a left- and a right-handed baryon number and a left- and right-handed unitary spin.

The situation is just as in Sec. III, where we defined a left- and a right-handed isotopic spin and we could have defined a left- and a right-handed nucleon number.

The left- and right-handed quantities are connected to each other by the parity operation P:

$$PF_i{}^\pm P^{-1} = F_i{}^\mp. \qquad (5.15)$$

Now that we have constructed the mathematical apparatus of the group $U(3) \times U(3)$ and its algebra, we may inquire how the Hamiltonian density H behaves under the group, i.e., under commutation with the algebra.

In the model, there is, corresponding to (4.6), the formula

$$H = \bar{H} - L' - L'', \qquad (5.16)$$

where $\bar{H}$ is the Hamiltonian density derived from the Lagrangian density $\bar{L}$ and is completely invariant under the group. Instead of defining u_0 as in Sec. III, let us put

$$u_0 = L' \propto \bar{b}\lambda_0 b. \qquad (5.17)$$

We can easily see that by commutation of u_0 with F_i and $F_i{}^5$ ($i = 0, \cdots, 8$) at equal times we obtain a set of eighteen quantities:

$$\begin{aligned} u_i &\propto \bar{b}\lambda_i b, \\ v_i &\propto -i\bar{b}\lambda_i\gamma_5 b. \end{aligned} \qquad (5.18)$$

In fact F_i acts like $\lambda_i/2$, $F_i{}^5$ like $\lambda_i\gamma_5/2$, u_i like $\beta\lambda_i$, and v_i like $-i\beta\gamma_5\lambda_i$. Thus we have at equal times[30]

$$\begin{aligned}{} [F_i, u_j] &= if_{ijk}u_k, \\ [F_i, v_j] &= if_{ijk}v_k, \\ [F_i{}^5, u_j] &= -id_{ijk}v_k, \\ [F_i{}^5, v_j] &= id_{ijk}u_k, \end{aligned} \qquad (5.19)$$

and the stronger relations

$$[\mathfrak{F}_{i4}(\mathbf{x},t), u_j(\mathbf{x}',t)] = -f_{ijk}u_k(\mathbf{x},t)\delta(\mathbf{x}-\mathbf{x}'), \text{ etc.} \quad (5.20)$$

for the densities. All indices run from 0 to 8.

Note that we can now express not only L' (which is defined to be u_0) but L'' as well, since by (4.7) it is proportional to u_8. We have, then,

$$H = \bar{H} - u_0 - cu_8, \qquad (5.21)$$

where c is of the order $(m_{0N} - m_{0\Lambda})/m_{0N}$ in the model.

We may now make a series of abstractions from the model. First, we suppose that currents $\mathfrak{F}_{i\alpha}$ and $\mathfrak{F}_{i\alpha}{}^5$ are defined, with commutation rules (5.3), (5.9), and (5.10), and with the weak current given by the analog of (5.5)[31]:

$$J_\alpha = \mathfrak{F}_{1\alpha} + \mathfrak{F}_{1\alpha}{}^5 + i\mathfrak{F}_{2\alpha} + i\mathfrak{F}_{2\alpha}{}^5$$
$$+ \mathfrak{F}_{4\alpha} + \mathfrak{F}_{4\alpha}{}^5 + i\mathfrak{F}_{5\alpha} + i\mathfrak{F}_{5\alpha}{}^5, \quad (5.22)$$

---

[28] Actually the Lagrangian (4.1) without the nucleon mass terms is invariant under a larger continuous group of transformations than the one $[U(3) \times U(3)]$ that we treat here. For example, there are infinitesimal transformations in which the baryon fields b acquire small terms in $\bar{b}$. Invariance under these is associated with the conservation of currents carrying baryon number 2. The author wishes to thank Professor W. Thirring for a discussion of these additional symmetries and of conformal transformations, which give still more symmetry.

[29] The groups $U(1)$, $SU(3)$, and $SU(2)$ cannot be further factored in this fashion. They are called *simple*.

[30] Note that even if we use just F_i and $F_i{}^5$ for $i = 1, \cdots, 8$, or $SU(3) \times SU(3)$ only, we still generate all eighteen u's and v's. [In the two-dimensional case described in Sec. III the situation is different. Using $SU(2) \times SU(2)$, we generate from u_0 only itself and v_1, v_2, v_3; if we then bring in $F_0{}^5$ as well, we obtain three more u's and one more v.] This remark is interesting because the group that gives currents known to be physically interesting is just $U(1) \times SU(3) \times SU(3)$; there is no known physical coupling to $\mathfrak{F}_{0\alpha}{}^5$, the axial vector baryon current.

[31] Note that the *total* weak current, whether for baryons and mesons or for leptons, is just a component of the current of an angular momentum. See reference 13.

while the electric current is given by (5.4). Next, we may take the Hamiltonian density to be of the form (5.21), with $\bar{H}$ invariant and u_i and v_i transforming as in (5.20). Then, if the theory is of the type described in Appendix A, we can calculate the divergences of the currents in terms of the equal-time commutators

$$\partial_\alpha \mathfrak{F}_{i\alpha} = i[F_i, u_0] + ic[F_i, u_8],$$
$$\partial_\alpha \mathfrak{F}_{i\alpha}{}^5 = i[F_i{}^5, u_0] + ic[F_i{}^5, u_8], \quad (5.23)$$

or, explicitly,

$$\partial_\alpha \mathfrak{F}_{i\alpha} = 0, \quad (i=0, 1, 2, 3, 8)$$
$$\partial_\alpha \mathfrak{F}_{4\alpha} = (\tfrac{2}{3})^{\frac{1}{2}} u_5, \text{ etc.,}$$
$$\partial_\alpha \mathfrak{F}_{0\alpha}{}^5 = (\tfrac{2}{3})^{\frac{1}{2}} v_0 + (\tfrac{2}{3})^{\frac{1}{2}} c v_8,$$
$$\partial_\alpha \mathfrak{F}_{1\alpha}{}^5 = [(\tfrac{2}{3})^{\frac{1}{2}} + (\tfrac{1}{3})^{\frac{1}{2}} c] v_1, \text{ etc.,} \quad (5.24)$$
$$\partial_\alpha \mathfrak{F}_{4\alpha}{}^5 = [(\tfrac{2}{3})^{\frac{1}{2}} - (\tfrac{1}{12})^{\frac{1}{2}} c] v_4, \text{ etc.,}$$
$$\partial_\alpha \mathfrak{F}_{8\alpha}{}^5 = [(\tfrac{2}{3})^{\frac{1}{2}} - (\tfrac{1}{3})^{\frac{1}{2}} c] v_8 + (\tfrac{2}{3})^{\frac{1}{2}} c v_0.$$

Finally, if we taken the model really seriously, we may abstract the equal-time commutation relations of the u_i and v_i as obtained by explicit commutation in the model.

The relations of Sec. III are all included in those of this section, except that what was called u_0 there is now called $u_0 + c u_8$ and what was called v_i is now called $[(\tfrac{2}{3})^{\frac{1}{2}} + (\tfrac{1}{3})^{\frac{1}{2}} c] v_i$ for $i = 1, 2, 3$.

All of the relations used here are supposed to be exact and are not affected by the symmetry-breaking character of the non-invariant term in the Hamiltonian. In the next section, we discuss what happens if c can be regarded as small in any sense. We may then expect to see some trace of the symmetry under $U(3)$ that would obtain if c were 0 and L'' disappeared. In this limit, N and Λ are degenerate, and all the components F_i of the unitary spin are conserved. The higher symmetry would show up particularly through the existence of degenerate baryon and meson supermultiplets, which break up into ordinary isotopic multiplets when L'' is turned on. These supermultiplets have been discussed previously for baryons and pseudoscalar mesons[4,6] and then for vector mesons.[32–34]

We shall not discuss the case in which both L' and L'' are turned off; that is the situation, still more remote from reality, in which all the axial vector currents are conserved as well as the vector ones.

VI. BROKEN SYMMETRY—MESON SUPERMULTIPLETS

We know that because of isotopic spin conservation the baryons and mesons form degenerate isotopic multiplets, each corresponding to an irreducible representation of the isotopic spin algebra (3.1). Each multiplet has $2I+1$ components, where the quantum num-

[32] M. Gell-Mann, California Institute of Technology Synchrotron Laboratory Report No. CTSL-20, 1961 (unpublished).
[33] Y. Ne'eman, Nuclear Phys. **26**, 222 (1961).
[34] A. Salam and J. C. Ward, Nuovo cimento **20**, 419 (1961).

ber I distinguishes one representation from another and gives us the eigenvalue $I(I+1)$ of the operator $\sum_{i=1}^{3} I_i^2$, which commutes with all the elements of the isotopic spin group. The operators I_i are represented, within the multiplet, by Hermitian $(2I+1) \times (2I+1)$ matrices having the commutation rules (3.1) of the algebra.

If we start from the doublet representation, we can build up all the others by considering combinations of particles that transform like the original doublet. Just as (p,n) form a doublet representation for which the I_i are represented by $\tau_i/2$, the antiparticles $(\bar{n}, -\bar{p})$ also form a doublet representation that is equivalent. (Notice the minus sign on the antiproton state or field.) Now, if we put together a nucleon and an antinucleon, we can form the combination

$$\bar{N} N = \bar{p} p + \bar{n} n,$$

which transforms like an isotopic singlet, or the combinations

$$\bar{N} \tau_i N, \quad (i=1, 2, 3)$$

which form an isotopic triplet. The direct product of nucleon and antinucleon doublets gives us a singlet and a triplet. Any meson that can dissociate virtually into nucleon and antinucleon must be either a singlet or a triplet. For the singlet state, the components I_i are all zero, while for the three triplet states the three 3×3 matrices, $I_i{}^{jk}$ of the components I_i, are given by

$$I_i{}^{jk} = -i e_{ijk}. \quad (6.1)$$

Now let us generalize these familiar results to the unitary spin and the three basic baryons b (comprising n, p, and Λ). These three fields or particles form a three-dimensional irreducible representation of the unitary spin algebra (4.16) from which all the other representations may be constructed.

For example, consider a meson that can dissociate into b and $\bar{b}$. It must transform either like

$$\bar{b} b = \bar{p} p + \bar{n} n + \bar{\Lambda} \Lambda,$$

a unitary singlet, or else like

$$\bar{b} \lambda_i b, \quad (i=1, \cdots, 8)$$

a unitary octet.

The unitary singlet is evidently neutral, with strangeness $S=0$, and forms an isotopic singlet. But how does the unitary octet behave with respect to isotopic spin? We form the combinations

$$\left. \begin{array}{l} \bar{b}(\lambda_1 - i\lambda_2) b / 2 = \bar{n} p, \\ \bar{b} \lambda_3 b / \sqrt{2} \quad = (\bar{p} p - \bar{n} n)/\sqrt{2}, \\ \bar{b}(\lambda_1 + i\lambda_2) b / 2 = \bar{p} n, \end{array} \right\} I=1, S=0$$

$$\left. \begin{array}{l} \bar{b}(\lambda_4 - i\lambda_5) b / 2 = \bar{\Lambda} p, \\ \bar{b}(\lambda_6 - i\lambda_7) b / 2 = \bar{\Lambda} n, \end{array} \right\} I=\tfrac{1}{2}, S=+1 \quad (6.2)$$

$$\left. \begin{array}{l} \bar{b}(\lambda_4 + i\lambda_5) b / 2 = \bar{p} \Lambda, \\ \bar{b}(\lambda_6 + i\lambda_7) b / 2 = \bar{n} \Lambda, \end{array} \right\} I=\tfrac{1}{2}, S=-1$$

$$\bar{b} \lambda_8 b / \sqrt{2} = (\bar{p} p + \bar{n} n - 2 \bar{\Lambda} \Lambda)/\sqrt{6}, \quad I=0, S=0,$$

and we see immediately that the unitary octet comprises an isotopic triplet with $S=0$, a pair of isotopic doublets with $S=\pm 1$, and an isotopic singlet with $S=0$. All these are degenerate only in the limit of unitary symmetry $(L''=0)$; when the mass-splitting term is turned on, the singlet, the triplet, and the pair of doublets should have three somewhat different masses.

The known pseudoscalar mesons (π, K, and $\bar{K}$) fit very well into this picture, provided there is an eighth pseudoscalar meson to fill out the octet. Let us call the hypothetical isotopic singlet pseudoscalar meson χ^0. Since it is pseudoscalar, it cannot dissociate (virtually or really) into 2π. It has the value $+1$ for the quantum number G, so that it cannot dissociate into an odd number of pions either. Thus in order to decay by means of the strong interactions, it must have enough energy to yield 4π. It would then appear as a 4π resonance. The decay into 4π is, however, severely hampered by centrifugal barriers.

If the mass of χ^0 is too low to permit it to decay readily into 4π, then it will decay electromagnetically. If there is sufficient energy, the decay mode $\chi^0 \to \pi^+ + \pi^- + \gamma$ is most favorable; otherwise[34a] it will decay into 2γ like π^0.

Let us now turn to the vector mesons. The best known vector meson is the $I=1$, $J=1^-$ resonance of 2π, which we shall call ρ. It has a mass of about 750 Mev. According to our scheme, it should belong, like the pion, to a unitary octet. Since it occupies the same position as the π ($I=1$, $S=0$), we denote it by the succeeding letter of the Greek alphabet.

The vector analog of χ^0 we shall call ω^0 (skipping the Greek letter ψ). It must have $I=0$, $J=1^-$, and $G=-1$ and so it is capable of dissociation into $\pi^+ + \pi^- + \pi^0$. Presumably it is the 3π resonance found experimentally[35] at about 790 Mev.

In order to complete the octet, we need a pair of strange doublets analogous to K and $\bar{K}$. In the vector case, we shall call them M and $\bar{M}$ (skipping the letter L). Now there is a known $K\pi$ resonance with $I=\frac{1}{2}$ at about 884 Mev. If it is a p-wave resonance, then it fits the description of M perfectly.

In the limit of unitary symmetry, we can have, besides the unitary octet of vector mesons, a unitary singlet. The hypothetical B^0 that we discussed in Sec. III would have such a character. If B^0 exists, then the turning-on of the mass-splitting term L'' mixes the states B^0 and ω^0, which are both *isotopic* singlets.

Other mesons may exist besides those discussed, for example, scalar and axial vector mesons. All those that can associate into $b+\bar{b}$ should form unitary octets or

---

[34a] *Note added in proof.* H. P. Duerr and W. Heisenberg (preprint) have pointed out the importance of the decay mode $\chi^0 \to 3\pi$ induced by electromagnetism. For certain χ masses, it may be a prominent mode.

[35] B. C. Maglić, L. W. Alvarez, A. H. Rosenfeld, and M. L. Stevenson, Phys. Rev. Letters **7**, 178 (1961).

TABLE III. Possible meson octets and singlets.

Unitary spin	Isotopic spin	Strangeness	Pseudoscalar	Vector	Scalar	Axial vector
Octet	1	0	π	ρ	π'	ρ'
	1/2	+1	K	M	K'	M'
	1/2	−1	$\bar{K}$	$\bar{M}$	$\bar{K}'$	$\bar{M}'$
	0	0	χ	ω	χ'	ω'
Singlet	0	0	A	B	A'	B'

singlets or both, with each octet splitting into isotopic multiplets because of the symmetry-breaking term L''.

A list of some possible meson states is given in Table III, along with suggested names for the mesons.

It is interesting that we can predict not only the degeneracy of an octet in the limit $L'' \to 0$ but also a sum rule[32] that holds in first order in L'':

$$(m_K+m_{\bar{K}})/2 = (3m_\chi + m_\pi)/4,$$
$$(m_M+m_{\bar{M}})/2 = (3m_\omega + m_\rho)/4. \quad (6.3)$$

If M is at about 884 Mev and ρ at about 750 Mev, then ω should lie at about 930 Mev according to the sum rule; since it is actually at 790 Mev, the sum rule does not seem to give a good description of the splitting. Perhaps an important effect is the repulsion between the ω^0 and B^0 levels, pushing ω^0 down and B^0 up. For what it is worth, (6.3) gives a χ^0 mass of around 610 Mev.

In the limit of unitary symmetry, not only are the supermultiplets degenerate but their effective couplings are symmetrical. For example, the effective coupling of the unitary pseudoscalar octet to N and Λ takes the form

$$ig_1\{\bar{N}\tau\gamma_5 N \cdot \pi + \bar{N}\gamma_5 \Lambda K + \bar{\Lambda}\gamma_5 N\bar{K} \\ + 3^{-\frac{1}{2}}\bar{N}\gamma_5 N\chi - 2\times 3^{-\frac{1}{2}}\bar{\Lambda}\gamma_5\Lambda\chi\}, \quad (6.4)$$

in terms of renormalized "fields." Now, as the term L'' is turned on, the various coupling constants become unequal; instead of calling them all g_1, we refer to them as $g_{NN\pi}$, g_{NAK}, $g_{NN\chi}$, and $g_{\Lambda\Lambda\chi}$, respectively, each of these constants being the measurable renormalized coupling parameter at the relevant pole.

We have written the effective coupling (6.4) as if there were renormalized fields for all the particles involved, but that is only a matter of notation; the mesons can perfectly well be composite. We may simplify the notation still further by constructing a traceless 3×3 matrix Π containing the pseudoscalar "fields" in such a way that (6.4) becomes

$$ig_1\bar{b}\Pi\gamma_5 b. \quad (6.5)$$

We may now write, in a trivial way, other effective couplings in the limit of unitary symmetry. We define a traceless 3×3 matrix W_α containing the "fields" for the vector meson octet just as Π^* contains those for the pseudoscalar octet. We then have the invariant effective coupling

$$i\gamma_1 \, \text{Tr} W_\alpha (\Pi \partial_\alpha \Pi - \partial_\alpha \Pi \Pi)/2 \qquad (6.6)$$

in the symmetric limit. When the asymmetry is turned on, the single coupling parameter γ_1 is replaced by the set of different parameters $\gamma_{\rho\pi\pi}$, $\gamma_{\rho KK}$, $\gamma_{\omega KK}$, $\gamma_{MK\pi}$, and $\gamma_{MK\chi}$.

In the same way, we have another effective coupling

$$ih_1 \, \text{Tr} \Pi (\partial_\alpha W_\beta - \partial_\beta W_\alpha)(\partial_\gamma W_\delta - \partial_\delta W_\gamma) e_{\alpha\beta\gamma\delta} \qquad (6.7)$$

in the symmetric limit; in the actual asymmetric case, we define the distinct constants $h_{\pi\omega\rho}$, $h_{\pi MM}$, $h_{\chi\omega\omega}$, $h_{\chi\rho\rho}$, $h_{\chi MM}$, $h_{KM\rho}$, and $h_{KM\omega}$. All of these constants can be measured, in principle, in "pole" experiments, except that for the broad resonances like ρ the poles are well off the physical sheet.

We have generalized the definitions of constants like $g_{NN\pi}$ and $\gamma_{\rho\pi\pi}$, as used in Sec. II, to other particles. The constants γ_ρ and f_π of Sec. II also have analogs, of course, and we define γ_ω, f_K, etc., in the obvious way. In the limit of unitary symmetry, of course, we would have $f_\pi = f_K = f_X$ and $\gamma_\rho = \gamma_\omega = \gamma_M$. Likewise, the constant $-G_A/G$ for nucleon β decay would equal the corresponding quantity $-G_A{}^{\Delta N}/G$ for the β decay of Λ.

VII. BROKEN SYMMETRY—BARYON SUPERMULTIPLETS

What has been done in the previous section may be described mathematically as follows. We considered a three-dimensional representation of the unitary spin algebra (4.16) or of the group $SU(3)$ that is generated by the algebra. It is the representation to which b belongs (that is, n, p, and Λ) and we may denote it by the symbol 3.

The antiparticles $\bar{b}$ belong to the conjugate representation 3^*, which is inequivalent[36] to 3. We have then taken the direct product $3 \times 3^*$ and found it to be given by the rule

$$3 \times 3^* = 8 + 1, \qquad (7.1)$$

where 8 is the octet representation and 1 the singlet representation of unitary spin. Each of these is its own conjugate; that is a situation that occurs only when the dimension is the cube of an integer.

There are, of course, more complicated representations to which mesons might belong that are incapable (in the limit of unitary symmetry) of dissociation into $b + \bar{b}$ but capable of dissociation into $2b + 2\bar{b}$ or higher configurations. But we might guess that at least the mesons of lowest mass would correspond to the lowest configurations.

Now we want to examine the simplest configurations

[36] In other words, no unitary transform can convert the representations 3 and 3* into each other. That is easy to see, since the eigenvalues of λ_8 are opposite in sign for the two representations, and changing the signs changes the set of eigenvalues. In the case of the group $SU(2)$ of isotopic spin transformations, the basic spinor representation $I = \frac{1}{2}$ *is* equivalent to the corresponding antiparticle representation.

for baryons, apart from just b. Evidently the next simplest is $2b + \bar{b}$, which poses the problem of reducing the direct product $3 \times 3 \times 3^*$; the result is the following:

$$3 \times 3 \times 3^* = 3 \times 1 + 3 \times 8 = 3 + 3 + 6 + 15. \qquad (7.2)$$

The six-dimensional representation **6** is composed of an isotopic triplet with $S = -1$, a doublet with $S = 0$, and a singlet with $S = +1$; the fifteen-dimensional representation **15** is composed of a doublet with $S = -2$, a singlet and a triplet with $S = -1$, a doublet and a quartet with $S = 0$, and a triplet with $S = +1$. According to the scheme, then, Ξ should belong to 15. Where are the other members of the supermultiplet? For $S = -1$ and $S = 0$, there are many known resonances, some of which might easily have the same spin and parity as Ξ. For $S = +1$, $I = 1$, however, no resonance has been found so far (in K^+-p scattering, for example).

The hyperon Σ should also be placed in a supermultiplet, which may or may not be the same one to which Ξ belongs; we do not know if the spin and parity of Σ and Ξ are the same, with K taken to be pseudoscalar. If Σ belongs to **6** in the limit of unitary symmetry, then there should be a KN resonance in the $I = 0$ state.

It is difficult to say at the present time if the baryon states can be reconciled with the model. Further knowledge of the baryon resonances is required.

One curious possibility is that the fundamental objects b are hidden and that the physical N and Λ, instead of belonging to 3, belong, along with Σ and Ξ, to the representation 15 in the limit of unitary symmetry. That would require the spins and parities of N, Λ, Σ, and Ξ to be equal, and it would require a πN resonance in the $p_{\frac{1}{2}}$, $I = \frac{3}{2}$ state as well as a KN resonance in the $p_{\frac{1}{2}}$, $I = 1$ state to fill out the supermultiplet.

VIII. THE "EIGHTFOLD WAY"

Unitary symmetry may be applied to the baryons in a more appealing way if we abandon the connection with the symmetrical Sakata model and treat unitary symmetry in the abstract. (An abstract approach is, of course, required if there are no "elementary" baryons and mesons.) Of all the groups that could be generated by the vector weak currents, $SU(3)$ is still the smallest and the one that most naturally gives rise to the rules $|\Delta I| = \frac{1}{2}$ and $\Delta S/\Delta Q = 0, +1$.

There is no longer any reason for the baryons to belong to the 3 representation or the other spinor representations of the group $SU(3)$; the various irreducible spinor representations are those obtained by reducing direct products like $3 \times 3 \times 3^*$, $3 \times 3 \times 3 \times 3^* \times 3^*$, etc.

Instead, the baryons may belong, like the mesons, to representations such as 8 or 1 obtained by reducing the direct products of equal numbers of 3's and 3^*'s. It is then natural to assign the stable and metastable baryons N, Λ, Σ, and Ξ to an octet, degenerate in the limit of unitary symmetry. We thus obtain the scheme of Gell-Mann[32] and Ne'eman[33] that we call the "eightfold way." The component F_8 of the unitary spin is now $(\sqrt{3}/2)Y$, where Y is the hypercharge (equal to strangeness plus baryon number).

The baryons of the octet must have the same spin and parity (treating K as pseudoscalar). To first order in the violation of unitary symmetry, the masses should obey the sum rule analogous to (6.3):

$$(m_N+m_\Xi)/2 = (3m_\Lambda+m_\Sigma)/4, \qquad (8.1)$$

which agrees surprisingly well with observations, the two sides differing by less than 20 Mev.

To form mesons that transform like combinations of these baryons and their antiparticles, we reduce the direct product 8×8 (remembering that $8=8^*$) and obtain

$$8\times 8 = 1+8+8+10+10^*+27, \qquad (8.2)$$

where 1 and 8 are the singlet and octet representations already discussed; 10 consists of an isotopic triplet with $Y=0$, a doublet with $Y=-1$, a quartet with $Y=+1$, and a singlet with $Y=-2$; 10^* has the opposite behavior with respect to Y; and 27 consists of an isotopic singlet, triplet, and quintet with $Y=0$, a pair of doublets with $Y=\pm 1$, a pair of quartets with $Y=\pm 1$, and a pair of triplets with $Y=\pm 2$. Evidently the known mesons are to be assigned to octets and perhaps singlets, as in Sec. VI. The meson-nucleon scattering resonances must then also be assigned representations among those in (8.2); the absence so far of any observed structure in K-N scattering makes it difficult to place the $I=3/2$, $J=3/2$, π-N resonance in a supermultiplet.

The fact that 8 occurs twice in Eq. (8.2) means that there are two possible forms of symmetrical Yukawa coupling of a meson octet to the baryon octet in the limit of unitary symmetry. As in Sec. VI for the mesons, we form a 3×3 traceless matrix out of the formal "fields" of the baryon octet; call it $\mathfrak{B}$. The effective symmetrical coupling of pseudoscalar mesons may then be written as

$$ig_1\alpha \, \mathrm{Tr}\, (\overline{\mathfrak{B}}\pi\gamma_5\mathfrak{B}+\pi\overline{\mathfrak{B}}\gamma_5\mathfrak{B})/2$$
$$+ig_1(1-\alpha)\, \mathrm{Tr}\, (\overline{\mathfrak{B}}\pi\gamma_5\mathfrak{B}-\pi\overline{\mathfrak{B}}\gamma_5\mathfrak{B})/2. \qquad (8.3)$$

The two types of coupling differ in their behavior under the operation R that exchanges N and Ξ, K and $\bar{K}$, M and $\bar{M}$, etc.; the first term is symmetric while the second is antisymmetric under R. The parameter α just specifies how much of each effective coupling is presented in the limit of unitary symmetry. When we take into account violations of the symmetry, we must define separate coupling constants $g_{NN\pi}$, $g_{NK\Lambda}$, etc., in a suitable way.

Likewise the vector mesons have the general symmetrical coupling

$$i\gamma_1\beta\, \mathrm{Tr}\, (\overline{\mathfrak{B}}W_\alpha\gamma_\alpha\mathfrak{B}+W_\alpha\overline{\mathfrak{B}}\gamma_\alpha\mathfrak{B})$$
$$+i\gamma_1(1-\beta)\, \mathrm{Tr}\, (\overline{\mathfrak{B}}W_\alpha\gamma_\alpha\mathfrak{B}-W_\alpha\overline{\mathfrak{B}}\gamma_\alpha\mathfrak{B}), \qquad (8.4)$$

where we ignore Pauli moment terms for simplicity. To the extent that the vector meson octet W_α dominates the dispersion relations for the unitary spin current $\mathfrak{F}_{i\alpha}$, then the mesons of W_α are coupled effectively to the components of $\mathfrak{F}_{i\alpha}$, and we have $\beta=0$ in (8.5). Then ρ is effectively coupled to the isotopic spin current, ω to the hypercharge current, and M to the strangeness-changing vector current. The first two of these currents are conserved, and so we have the approximate universality of ρ and ω couplings proposed by Sakurai[11] and discussed in Sec. II. In the limit of unitary symmetry, under the assumptions just mentioned, ρ is effectively coupled to the current of $2\gamma_1\mathbf{I}$ and ω to the current of $2\gamma_1F_8=\sqrt{3}\gamma_1Y$.

The electromagnetic current is now given by the formula

$$j_\alpha = \mathfrak{F}_{3\alpha}+3^{-\frac{1}{2}}\mathfrak{F}_{8\alpha} \qquad (8.5)$$

instead of (5.4), while the weak vector current is still described by Eq. (5.5). If we are to treat the vector and axial vector currents by means of $SU(3)\times SU(3)$, as we did earlier, then the entire weak current is given by (5.22) and we have the commutation rules (5.3), (5.9), and (5.10) for the various components of the currents. The question of the behavior of H under the group $SU(3)\times SU(3)$ should, however, be re-examined for the eightfold way; we shall not go into that question here. But let us consider how the baryon octet transforms in the limit of conserved vector *and* axial vector currents [invariance under $SU(3)\times SU(3)$]. In the Sakata model, the left-handed baryons transformed under (F_i^+,F_j^-) like $(3, 1)$, while the right-handed baryons transformed according to $(1, 3)$. For the eightfold way, there are two simple possibilities for these transformation properties. Either we have $(8, 1)$ and $(1, 8)$ or else we adjoin a ninth neutral baryon (which need not be degenerate with the other eight in the limit of conserved *vector* currents and which need not have the same parity) and use the transformation properties $(3, 3^*)$ and $(3^*, 3)$. In the first case, the baryons transform like the quantities $\mathfrak{F}_{i\alpha}$ and $\mathfrak{F}_{i\alpha}^5$ $(i=1,\cdots,8)$ and in the second case they transform like u_i and v_i $(i=0,\cdots,8)$ of Sec. V.

IX. REMARKS AND SUGGESTIONS

Our approach to the problem of baryon and meson couplings leads to a number of suggestions for new investigations, both theoretical and experimental.

First, the equal-time commutation relations for currents and densities lead to exact sum rules for the weak and electromagnetic matrix elements. As an example, take the commutation rules (3.5) for the isotopic spin current. These do not, of course, depend on any higher symmetry, but they can be used to illustrate the results that can be obtained from the more general relations like (5.3).

Consider the electromagnetic form factor $F_\pi(s)$ of the charged pion, which is just the form factor of the isotopic spin current between one-pion states. Let p and p' be the initial and final pion four-momenta, with $s=-(p-p')^2$. Let K be any four-momentum with $K^2=-m_\pi^2$. Then, taking the matrix element of (3.5) between one-pion states, we obtain the result

$$2(p_0+p_0')K_0F_\pi(-(p-p')^2)$$
$$=(p_0+K_0)(p_0'+K_0)F_\pi(-(p-K)^2)F_\pi(-(p'-K)^2)$$
$$-(p_0-K_0)(p_0'-K_0)F_\pi(-(p+K)^2)F_\pi(-(p'+K)^2)$$
$$+\text{inelastic terms}, \quad (9.1)$$

where the inelastic terms come from summing over bilinear forms in the inelastic matrix elements of the current. We see that if there is no inelasticity the form factor is unity. Thus the departure from unity of $F_\pi(s)$ is related to the amount of inelasticity.

A similar relation is familiar in nonrelativistic quantum mechanics:

$$\langle e^{i(p-p')\cdot x}\rangle_{00} = \sum_n \langle e^{i(p-k)\cdot x}\rangle_{0n} \langle e^{i(k-p')\cdot x}\rangle_{n0}. \quad (9.2)$$

If we apply relations like (9.1) to the matrix elements of non-conserved currents like $\mathbf{P}_\alpha$, along with the linear homogeneous dispersion relations for these matrix elements, we can in principle determine constants like $-G_A/G$.

A second line of theoretical investigation is suggested by the vanishing at high momentum transfer of matrix elements of divergences of non-conserved currents, like $\partial_\alpha \mathbf{P}_\alpha$. We should try to find limits involving high energies and high momentum transfers in which we can show that the conservation of helicity, unitary spin, etc., becomes valid. A preliminary effort in this direction has been made by Gell-Mann and Zachariasen.[37]

A third topic of study is the testing of broken symmetry at low energy. Do the mesons fall into unitary octets and singlets? An experimental search for χ^0 is required and also a determination of the spin of K^* at 884 Mev to see if it really is our M meson.

Let us discuss briefly the properties of χ^0. An $I=0$ state of 4π can have two types of symmetry: either totally symmetric (partition [4]) in both space and isotopic spin or else the symmetry of the partition [2+2] in space and in isotopic spin. For a pseudoscalar state, the first type of wave function in momentum space is very complicated. If $\mathbf{p}$, $\mathbf{q}$, and $\mathbf{r}$ are the three momentum differences, it must look like

$$\mathbf{p} \cdot \mathbf{q} \times \mathbf{r}(E_1-E_2)(E_2-E_3)(E_3-E_4)$$
$$\times (E_1-E_3)(E_1-E_4)(E_2-E_4),$$

times a symmetric function of the energies E_1, E_2, E_3, E_4 of the four pions. On the basis of any reasonable dynamical picture of χ^0, such a wave function should have a very small amplitude. In particular, dispersion theory suggests that the wave function of χ^0 should have large contributions from virtual dissociation into 2ρ, which gives a wave function with [2+2] symmetry.

[37] M. Gell-Mann and F. Zachariasen, Phys. Rev. 123, 1065 (1961).

If [2+2] predominates, then the charge ratio in decay is 2:1 in favor of $2\pi^0+\pi^++\pi^-$ over $2\pi^++2\pi^-$, with $4\pi^0$ absent. If virtual dissociation into 2ρ actually predominates, then the matrix element of the 4π configuration is easily written down and the spectrum of the decay $\chi^0 \to 4\pi$ can be calculated.

If χ^0 is lighter than 4π, it will, of course, decay electromagnetically. Even if it is above threshold for 4π, however, the matrix element for decay contains so many powers of pion momenta that electromagnetic decay should be appreciable over a large range of masses. The branching ratio $(\pi^++\pi^-+\gamma)/(4\pi)$ is approximately calculable by dispersion methods. In both cases χ^0 first dissociates into 2ρ. Then either both virtual ρ mesons decay into 2π, or else (in the case where both are neutral) one may decay into $\pi^++\pi^-$, while the other turns directly into γ. If we draw a diagram for such a process, then the constant $\gamma_{\rho\pi\pi}$ is inserted whenever we have a $\rho\pi\pi$ vertex and the constant $em_\rho^2/2\gamma_\rho$ at a ρ-γ junction.[10]

If the meson spectrum is consistent with broken unitary symmetry, we should examine the baryons, and see whether the various baryon states fit into the representations $\mathbf{3}$, $\mathbf{6}$, and $\mathbf{15}$ (or the representations $\mathbf{1}$, $\mathbf{8}$, $\mathbf{10}$, $\mathbf{10}^*$, and $\mathbf{27}$ that arise in the alternative form of unitary symmetry).

If some states are lacking in a given supermultiplet, it does not necessarily prove that the broken symmetry is wrong, but only that it is badly violated. We assume that baryon isobars like the $\pi N \frac{3}{2}, \frac{3}{2}$ resonance are dynamical in nature; there may be some attractive and some repulsive forces in this channel, and the attractive ones have won out, producing the resonance. In the KN channel with $I=1$, for example, it is conceivable that the repulsive ones are stronger (because of symmetry violation), and the analogous $p_\frac{3}{2}$ resonance disappears. In such a situation, the concept of broken symmetry at low energies is evidently of little value.

Suppose, however, that the idea of broken unitary symmetry is confirmed for both mesons and baryons, say according to the Sakata picture, in which N and Λ belong to the representation $\mathbf{3}$ in the limit of unitary symmetry. There are, nevertheless, gross violations of unitary symmetry, and the elucidation of these, both theoretical and experimental, is a fourth interesting subject.

If unitary symmetry were exact, then not only would m_K/m_π equal unity, instead of about 3.5, but $f_K{}^2/f_\pi{}^2$ would be 1 instead of about 6, and $3G_A{}^2+G_V{}^2$ for the β decay of Λ would be equal to $3G_A{}^2+G_V{}^2$ for the nucleon instead of being 1/15 as large. All these huge departures from unity represent very serious violations of unitary symmetry.

Yet the relatively small mass difference of N and Λ compared to their masses would seem to indicate, if our model is right, that the constant c in Eq. (5.21) is considerably smaller than unity. It is conceivable that the large mass ratio of K to π comes about because the total mass of the system is so small. It is possible that even with a fairly small c (say $\sim -\frac{1}{10}$) we might explain the gross violations of unitary symmetry. We might try to interpret the large values of $g_{NN\pi}{}^2/g_{NK\Lambda}{}^2$, $f_K{}^2/f_\pi{}^2$, etc., in terms of the large value of $m_K{}^2/m_\pi{}^2$.

An example of such a calculation, and one that illustrates the various methods suggested in this article, is the following. We try to calculate $f_K{}^2/f_\pi{}^2$ in terms of $m_K{}^2/m_\pi{}^2$.

Consider the following vacuum expectation value, written in parametric representation:

$$\langle [\mathcal{F}_{1\alpha}{}^5(x),\partial_\beta\mathcal{F}_{1\beta}{}^5(x')]\rangle_0 = i/(2\pi)^3 \int d^4K\, e^{iK\cdot(x-x')}$$
$$\times K_\alpha \epsilon(K) \int dM^2/M^2 \delta(K^2+M^2)\rho(M^2). \quad (9.3)$$

Here x and x' are arbitrary space-time points. In terms of (9.3), we have

$$\langle [\partial_\alpha\mathcal{F}_{1\alpha}{}^5(x),\partial_\beta\mathcal{F}_{1\beta}{}^5(x')]\rangle = 1/(2\pi)^3 \int d^4K\, e^{iK\cdot(x-x')}$$
$$\times \epsilon(K) \int dM^2\, \delta(K^2+M^2)\rho(M^2). \quad (9.4)$$

Now the contribution of the one-pion intermediate state is easily obtained in terms of the constant $f_\pi{}^2$:

$$\rho(M^2) = \delta(M^2-m_\pi{}^2)m_\pi{}^4/4f_\pi{}^2 + \text{higher terms}. \quad (9.5)$$

If $\int \rho(M^2)dM^2/M^2$ converges and if the one-pion term dominates, we have

$$\int \rho(M^2)dM^2/M^2 \approx m_\pi{}^2/4f_\pi{}^2. \quad (9.6)$$

But from (9.3) we can extract the expectation value of the equal-time commutator of the fourth component of $\mathcal{F}_{1\alpha}{}^5$ with $\partial_\beta\mathcal{F}_{1\beta}{}^5$; making use of (5.20) and (5.24), we can express the result in terms of $\langle u_0\rangle$ and $\langle u_8\rangle$. Thus we find

$$\int \rho(M^2)dM^2/M^2 = [(2/3)^{\frac{1}{2}}+(1/3)^{\frac{1}{2}}c]$$
$$\times [(2/3)^{\frac{1}{2}}\langle u_0\rangle_0 + (1/3)^{\frac{1}{2}}\langle u_8\rangle_0], \quad (9.7)$$

assuming convergence.

Now we can do exactly the same thing for $\mathcal{F}_{4\alpha}{}^5$ and the K meson, obtaining, in place of the formula

$$m_\pi{}^2/4f_\pi{}^2 \approx [(\tfrac{2}{3})^{\frac{1}{2}}+(\tfrac{1}{3})^{\frac{1}{2}}c][(\tfrac{2}{3})^{\frac{1}{2}}\langle u_0\rangle_0 + (\tfrac{1}{3})^{\frac{1}{2}}\langle u_8\rangle_0], \quad (9.8)$$

the analogous result

$$m_K{}^2/4f_K{}^2 \approx [(\tfrac{2}{3})^{\frac{1}{2}} - (\tfrac{1}{12})^{\frac{1}{2}}c][(\tfrac{2}{3})^{\frac{1}{2}}\langle u_0\rangle_0 - (\tfrac{1}{12})^{\frac{1}{2}}\langle u_8\rangle_0]. \quad (9.9)$$

If c is really small, presumably $\langle u_8\rangle_0$ is also small compared to $\langle u_0\rangle_0$. Then we can, roughly, set (9.8) equal to (9.9), obtaining

$$f_K{}^2/f_\pi{}^2 \approx m_K{}^2/m_\pi{}^2. \quad (9.10)$$

The left-hand side is about 6 and the right-hand side about 10. Thus we can, in a crude approximation, calculate the rate of $K^+ \to \mu^+ + \nu$ in terms of that for $\pi^+ \to \mu^+ + \nu$ and explain one large violation of symmetry in terms of another.

The Goldberger-Treiman formula relating f_π, $g_{NN\pi}$, and $(-G_A/G)$ can also be used for the K particle to give a relation among f_K, $g_{N\Lambda K}$, and $(-G_A/G)$ for the β decay of Λ. Of course, the K-particle pole is much closer to the branch line beginning at $(m_K+2m_\pi)^2$ than the pion pole is to the branch line beginning at $9m_\pi{}^2$; thus the Goldberger-Treiman formula may be quite bad for the K meson. Still, we may try to use it to discuss the coupling of N and Λ to K and to leptons. We have

$$(m_N+m_\Lambda)(-G_A{}^{N\Lambda}/G) \approx g_{N\Lambda K}/f_K, \quad (9.11)$$

by analogy with (2.8). Comparing the two formulas, we have

$$(-G_A{}^{N\Lambda}/G)^2(-G_A/G)^{-2}$$
$$\approx g_{N\Lambda K}{}^2 g_{NN\pi}{}^{-2}(2m_Nf_\pi)^2[(m_\Lambda+m_N)f_K]^{-2}. \quad (9.12)$$

The ratio of g^2 factors is thought to be ~ 0.1 from photoproduction of K, while the remaining factor on the right is also ~ 0.1, so that the Goldberger-Treiman relation leads us to expect a very small axial vector β-decay rate for the Λ, much smaller than the observed one. The observed β decay would be nearly all vector; this prediction of the Goldberger-Treiman formula can easily be checked by observing the electron-neutrino angular correlation in the β decay of Λ, using bubble chambers.

We should, of course, try to predict the value of $g_{N\Lambda K}{}^2 g_{NN\pi}{}^{-2}$ in terms of $m_K{}^2/m_\pi{}^2$ just as we did above for $f_K{}^2/f_\pi{}^2$; however, it is a much harder problem.

When we know more about the coupling constants of the vector mesons (strong coupling constants such as $\gamma_{\omega NN}$, $h_{\omega\pi\rho}$, etc., and coupling strengths of currents such as γ_ω, γ_M, etc.) we will be able to make a survey of the pattern of coupling constants as well as the pattern of masses and see whether the higher symmetry has any relevance. Also it should become clear how well the approximation of dominant low-mass states works, in terms of universality of meson couplings and Goldberger-Treiman relations.[28]

In summary, then, we suggest the use of the equal-time commutators to predict sum rules, attempts to derive high-energy conservation laws and to check them

---

[28] An interesting relation of the Goldberger-Treiman type is one that holds if the trace $\theta_{\alpha\alpha}$ of the stress-energy-momentum tensor has matrix elements obeying highly convergent dispersion relations. Because of the vanishing of the self-stress, the expectation value of $\theta_{\alpha\alpha}$ in the state of a particle at rest gives the mass of the particle. Rewriting the matrix element as one between the vacuum and a one-pair state, we see that the dispersion relation involves intermediate states with $I=0$, $J=0^+$, $G=+1$. If there is a resonance or quasi-resonance in this channel (like the χ' meson of Table III) and if that resonance dominates the dispersion relation at low momentum transfers, then the coupling of the resonant state to different particles is roughly proportional to their masses. That is just the situation discussed by Schwinger in reference 1 and by Gell-Mann and Lévy in reference 17 for the "σ meson."

experimentally, the search for broken symmetry at low energies, attempts to calculate some violations in terms of others, and efforts to check the highly convergent dispersion relations dominated by low-mass states.

Nowhere does our work conflict with the program of Chew et al. of dynamical calculation of the S matrix for strong interactions, using dispersion relations. If something like the Sakata model is correct, then most of the mesons are dynamical bound states or resonances, and their properties are calculable according to the program. Those particles for which there are fundamental fields (like n, p, Λ, and B^0 in the specific field-theoretic model) would presumably occur as CDD poles or resonances in the dispersion relations.[39]

If there are no fundamental fields and no CDD poles, all baryons and mesons being bound or resonant states of one another, models like that of Sakata will fail; the symmetry properties that we have abstracted can still be correct, however. This situation would presumably differ in two ways[10] from the one mentioned above. First, all the masses and coupling constants could be calculated from coupled dispersion relations. Second, certain scattering amplitudes at high energies would show different behavior, corresponding to different kinds of subtractions in the dispersion relations. The second point should be investigated further, as it could lead to experimental tests of the "fundamental" character of various particles.[10,40]

ACKNOWLEDGMENTS

It is a pleasure to thank R. P. Feynman, S. L. Glashow, and R. Block for many stimulating discussions of symmetry, and to acknowledge the great value of conversations with G. F. Chew, S. Frautschi, R. Haag, R. Schroer, and F. Zachariasen about the explanation of approximate universality in terms of highly convergent dispersion relations.

APPENDIX

The field theories of the Fermi-Yang and Sakata models, given by Eqs. (3.17) and (4.1), respectively, belong to a general class of theories, which we now describe.

The Lagrangian density L is given as a function of a number of fields ψ_A and their gradients. The kinetic part of the Lagrangian (consisting of those terms containing gradients) is invariant under a set of infinitesimal unitary transformations generated by N independent Hermitian operators R_i, which may depend on the time. Under the transformations, the various fields ψ_A undergo linear recombinations:

$$\psi_A(\mathbf{x},t) \to \psi_A(\mathbf{x},t) - i\Lambda_i[R_i(t),\psi_A(\mathbf{x},t)]$$
$$= \psi_A(\mathbf{x},t) + i\Lambda_i \sum_B M_i^{AB}\psi_B(\mathbf{x},t), \quad \text{(A1)}$$

[39] L. Castillejo, R. H. Dalitz, and F. J. Dyson, Phys. Rev. **101**, 453 (1956).
[40] S. C. Frantschi, M. Gell-Mann, and F. Zachariasen (to be published).

where Λ_i is the infinitesimal gauge constant associated with the ith transformation. The equal-time commutation rules of the R_i are the same as those of the matrices M_i. Moreover, the set of R_i and linear combinations of R_i is algebraically complete under commutation; in other words, we have an algebra. The matrices M_i are the basis of a representation of the algebra (in general, a reducible representation). It is convenient to take the matrices of the basis to be orthonormal,

$$\text{Tr}M_iM_j = (\text{const})\delta_{ij}, \quad \text{(A2)}$$

redefining the R_i accordingly. The structure constants c_{ijk} in the commutation rules

$$[M_i,M_j] = ic_{ijk}M_k,$$
$$[R_i(t),R_j(t)] = ic_{ijk}R_k(t), \quad \text{(A3)}$$

are now real and totally antisymmetric in i, j, and k. We may still perform real rotations in the N-dimensional space of the R_i or the M_i. Suppose, after performing such a rotation, that we can split the R_i into two sets that commute with each other. Then our algebra is the direct sum of two commuting algebras. We continue this process until no further splitting is possible, even after performing rotations. The algebra has then been expressed as the direct sum of *simple* algebras. All the simple algebras have been listed by Cartan.[40] Besides the trivial one-dimensional algebra of $U(1)$ (which is not included by the mathematicians), there are the three-dimensional algebra of $SU(2)$, the eight-dimensional algebra of $SU(3)$, and so forth.

Now let us construct the currents of the operators R_i. We consider the gauge transformation of the second kind

$$\psi_A \to \psi_A(\mathbf{x},t) - i\Lambda_i(\mathbf{x},t)[R_i(t),\psi_A(\mathbf{x},t)], \quad \text{(A4)}$$

and ask what change it induces in the Lagrange density L. There will be a term in Λ_i and a term in $\partial_\alpha \Lambda_i$, so adjusted[17] that the total change is just the divergence of a four-vector:

$$L \to L(\mathbf{x},t) - \partial_\alpha \Lambda_i(\mathbf{x},t) R_{i\alpha}(\mathbf{x},t) - \Lambda_i(\mathbf{x},t) \partial_\alpha R_{i\alpha}(\mathbf{x},t). \quad \text{(A5)}$$

We define $R_{i\alpha}$ to be the current of R_i. It can be shown that R_i is in fact given by the relation

$$R_i = -i \int R_{i4} d^3x. \quad \text{(A6)}$$

Now if, for constant Λ_i, the whole Lagrangian is invariant under R_i, then the term in Λ_i in (A4) must vanish; we have $\partial_\alpha R_{i\alpha} = 0$. In other words, if there is exact symmetry under R_i, the current $R_{i\alpha}$ is conserved.

If there is a noninvariant part of L with respect to the symmetry operation R_i, then the current will not be conserved. By hypothesis, the noninvariant term (call it u) contains no gradients. Therefore, the effect

[40] E. Cartan, *Sur la Structure des groupes de transformations finis et continus*, thèse (Paris, 1894; 2nd ed., 1933).

of the transformation (A3) for *constant* Λ_i will be simply to add a term $-i\Lambda_i[R_i,u]$ to the Lagrangian density. We have, then, using (A4), the result

$$\partial_\alpha R_{i\alpha}(\mathbf{x},t) = i[R_i(t), u(\mathbf{x},t)]. \quad (A7)$$

Since u contains no gradients, it is not only the noninvariant term in the Lagrangian density, but also the negative of the noninvariant term in the Hamiltonian density. The invariant part of H evidently commutes with R_i. Thus we have

$$\partial_\alpha R_{i\alpha}(\mathbf{x},t) = -i[R_i(t), H(\mathbf{x},t)]. \quad (A8)$$

By considering the transformation properties of H under commutation with the algebra, we generate the divergences of all the currents. The formula obtained by integrating (A6) over space is, of course, very familiar:

$$\dot{R}_i = \int \partial_\alpha R_{i\alpha} d^3x = -i\left[R_i, \int H d^3x\right]. \quad (A9)$$

논문 웹페이지

Discovery of a Narrow Resonance in e^+e^- Annihilation*

J.-E. Augustin,† A. M. Boyarski, M. Breidenbach, F. Bulos, J. T. Dakin, G. J. Feldman,
G. E. Fischer, D. Fryberger, G. Hanson, B. Jean-Marie,† R. R. Larsen, V. Lüth,
H. L. Lynch, D. Lyon, C. C. Morehouse, J. M. Paterson, M. L. Perl,
B. Richter, P. Rapidis, R. F. Schwitters, W. M. Tanenbaum,
and F. Vannucci‡

Stanford Linear Accelerator Center, Stanford University, Stanford, California 94305

and

G. S. Abrams, D. Briggs, W. Chinowsky, C. E. Friedberg, G. Goldhaber, R. J. Hollebeek,
J. A. Kadyk, B. Lulu, F. Pierre,§ G. H. Trilling, J. S. Whitaker,
J. Wiss, and J. E. Zipse

Lawrence Berkeley Laboratory and Department of Physics, University of California, Berkeley, California 94720

(Received 13 November 1974)

We have observed a very sharp peak in the cross section for $e^+e^- \to$ hadrons, e^+e^-, and possibly $\mu^+\mu^-$ at a center-of-mass energy of 3.105 ± 0.003 GeV. The upper limit to the full width at half-maximum is 1.3 MeV.

We have observed a very sharp peak in the cross section for $e^+e^- \to$ hadrons, e^+e^-, and possibly $\mu^+\mu^-$ in the Stanford Linear Accelerator Center (SLAC)–Lawrence Berkeley Laboratory magnetic detector[1] at the SLAC electron-positron storage ring SPEAR. The resonance has the parameters

$E = 3.105 \pm 0.003$ GeV,

$\Gamma \leq 1.3$ MeV

(full width at half-maximum), where the uncertainty in the energy of the resonance reflects the uncertainty in the absolute energy calibration of the storage ring. [We suggest naming this structure $\psi(3105)$.] The cross section for hadron production at the peak of the resonance is ≥ 2300 nb, an enhancement of about 100 times the cross section outside the resonance. The large mass, large cross section, and narrow width of this structure are entirely unexpected.

Our attention was first drawn to the possibility of structure in the $e^+e^- \to$ hadron cross section during a scan of the cross section carried out in 200-MeV steps. A 30% (6 nb) enhancement was observed at a c.m. energy of 3.2 GeV. Subsequently, we repeated the measurement at 3.2 GeV and also made measurements at 3.1 and 3.3 GeV. The 3.2-GeV results reproduced, the 3.3-GeV measurement showed no enhancement, but the 3.1-GeV measurements were internally inconsistent—six out of eight runs giving a low cross section and two runs giving a factor of 3 to 5 higher cross section. This pattern could have been caused by a very narrow resonance at an energy slightly larger than the nominal 3.1-GeV setting of the storage ring, the inconsistent 3.1-GeV cross sections then being caused by setting errors in the ring energy. The 3.2-GeV enhancement would arise from radiative corrections which give a high-energy tail to the structure.

We have now repeated the measurements using much finer energy steps and using a nuclear magnetic resonance magnetometer to monitor the ring energy. The magnetometer, coupled with measurements of the circulating beam position in the storage ring made at sixteen points around the orbit, allowed the relative energy to be determined to 1 part in 10^4. The determination of the absolute energy setting of the ring requires the knowledge of $\int B\,dl$ around the orbit and is accurate to $\pm 0.1\%$.

The data are shown in Fig. 1. All cross sections are normalized to Bhabha scattering at 20 mrad. The cross section for the production of hadrons is shown in Fig. 1(a). Hadronic events are required to have in the final state either ≥ 3 detected charged particles or 2 charged particles noncoplanar by $> 20°$.[2] The observed cross section rises sharply from a level of about 25 nb to a value of 2300 ± 200 nb at the peak[3] and then exhibits the long high-energy tail characteristic of radiative corrections in e^+e^- reactions. The detection efficiency for hadronic events is 45% over the region shown. The error quoted above includes both the statistical error and a 7% contribution from uncertainty in the detection efficiency.

Our mass resolution is determined by the energy spread in the colliding beams which arises from quantum fluctuations in the synchrotron radiation emitted by the beams. The expected Gaussian c.m. energy distribution ($\sigma = 0.56$ MeV), folded with the radiative processes,[4] is shown as the dashed curve in Fig. 1(a). The width of the resonance must be smaller than this spread; thus an upper limit to the full width at half-maximum is 1.3 MeV.

Figure 1(b) shows the cross section for e^+e^- final states. Outside the peak this cross section

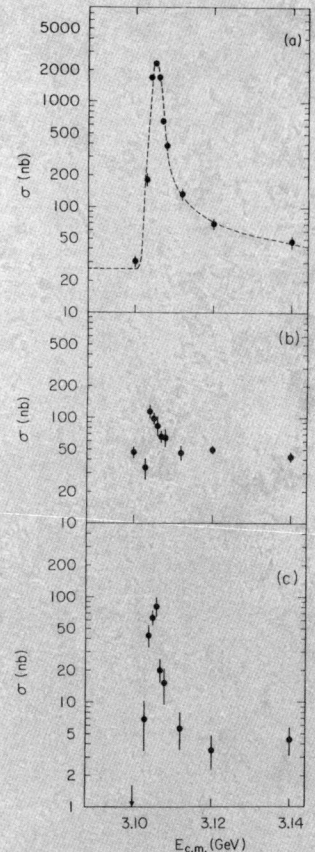

FIG. 1. Cross section versus energy for (a) multihadron final states, (b) e^+e^- final states, and (c) $\mu^+\mu^-$, $\pi^+\pi^-$, and K^+K^- final states. The curve in (a) is the expected shape of a δ-function resonance folded with the Gaussian energy spread of the beams and including radiative processes. The cross sections shown in (b) and (c) are integrated over the detector acceptance. The total hadron cross section, (a), has been corrected for detection efficiency.

is equal to the Bhabha cross section integrated over the acceptance of the apparatus.[1]

Figure 1(c) shows the cross section for the production of collinear pairs of particles, excluding electrons. At present, our muon identifications system is not functioning and we therefore cannot separate muons from strongly interacting particles. However, outside the peak the data are consistent with our previously measured μ-pair cross section. Since a large $\pi\pi$ or KK branching ratio would be unexpected for a resonance this massive, the two-body enhancement observed is *probably* but not *conclusively* in the μ-pair channel.

The $e^+e^- \to$ hadron cross section is presumed to go through the one-photon intermediate state with angular momentum, parity, and charge conjugation quantum numbers $J^{PC} = 1^{--}$. It is difficult to understand how, without involving new quantum numbers or selection rules, a resonance in this state which decays to hadrons could be so narrow.

We wish to thank the SPEAR operations staff for providing the stable conditions of machine performance necessary for this experiment. Special monitoring and control techniques were developed on very short notice and performed excellently.

---

*Work supported by the U. S. Atomic Energy Commission.

†Present address: Laboratoire de l'Accélérateur Linéaire, Centre d'Orsay de l'Université de Paris, 91 Orsay, France.

‡Permanent address: Institut de Physique Nucléaire, Orsay, France.

§Permanent address: Centre d'Etudes Nucléaires de Saclay, Saclay, France.

[1]The apparatus is described by J.-E. Augustin *et al.*, to be published.

[2]The detection-efficiency determination will be described in a future publication.

[3]While preparing this manuscript we were informed that the Massachusetts Institute of Technology group studying the reaction $pp \to e^+e^- + x$ at Brookhaven National Laboratory has observed an enhancement in the e^+e^- mass distribution at about 3100 MeV. J. J. Aubert *et al.*, preceding Letter [Phys. Rev. Lett. 33, 1402 (1974)].

[4]G. Bonneau and F. Martin, Nucl. Phys. B27, 381 (1971).

논문 웹페이지

Experimental Observation of a Heavy Particle J†

J. J. Aubert, U. Becker, P. J. Biggs, J. Burger, M. Chen, G. Everhart, P. Goldhagen,
J. Leong, T. McCorriston, T. G. Rhoades, M. Rohde, Samuel C. C. Ting, and Sau Lan Wu
Laboratory for Nuclear Science and Department of Physics, Massachusetts Institute of Technology, Cambridge, Massachusetts 02139

and

Y. Y. Lee
Brookhaven National Laboratory, Upton, New York 11973
(Received 12 November 1974)

We report the observation of a heavy particle J, with mass $m = 3.1$ GeV and width approximately zero. The observation was made from the reaction $p + \text{Be} \rightarrow e^+ + e^- + x$ by measuring the e^+e^- mass spectrum with a precise pair spectrometer at the Brookhaven National Laboratory's 30-GeV alternating-gradient synchrotron.

This experiment is part of a large program to study the behavior of timelike photons in $p + p \rightarrow e^+ + e^- + x$ reactions[1] and to search for new particles which decay into e^+e^- and $\mu^+\mu^-$ pairs.

We use a slow extracted beam from the Brookhaven National Laboratory's alternating-gradient synchrotron. The beam intensity varies from 10^{10} to 2×10^{12} p/pulse. The beam is guided onto an extended target, normally nine pieces of 70-mil Be, to enable us to reject the pair accidentals by requiring the two tracks to come from the same origin. The beam intensity is monitored with a secondary emission counter, calibrated daily with a thin Al foil. The beam spot size is 3×6 mm², and is monitored with closed-circuit television. Figure 1(a) shows the simplified side view of one arm of the spectrometer. The two arms are placed at 14.6° with respect to the incident beam; bending (by $M1$, $M2$) is done vertically to decouple the angle (θ) and the momentum (p) of the particle.

The Cherenkov counter C_0 is filled with one atmosphere and C_e with 0.8 atmosphere of H_2. The counters C_0 and C_e are decoupled by magnets $M1$ and $M2$. This enables us to reject knock-on electrons from C_0. Extensive and repeated calibra-

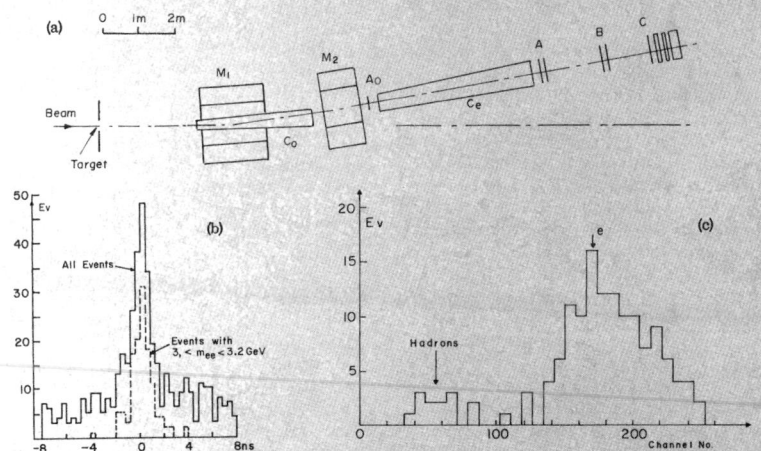

FIG. 1. (a) Simplified side view of one of the spectrometer arms. (b) Time-of-flight spectrum of e^+e^- pairs and of those events with $3.0 < m < 3.2$ GeV. (c) Pulse-height spectrum of e^- (same for e^+) of the e^+e^- pair.

tion of all the counters is done with approximately 6-GeV electrons produced with a lead converter target. There are eleven planes ($2 \times A_0$, $3 \times A$, $3 \times B$, $3 \times C$) of proportional chambers rotated approximately 20° with respect to each other to reduce multitrack confusion. To further reduce the problem of operating the chambers at high rate, eight vertical and eight horizontal hodoscope counters are placed behind chambers A and B. Behind the largest chamber C (1 m × 1 m) there are two banks of 25 lead glass counters of 3 radiation lengths each, followed by one bank of lead-Lucite counters to further reject hadrons from electrons and to improve track identification. During the experiment all the counters are monitored with a PDP 11-45 computer and all high voltages are checked every 30 min.

The magnets were measured with a three-dimensional Hall probe. A total of 10^5 points were mapped at various current settings. The acceptance of the spectrometer is $\Delta\theta = \pm 1°$, $\Delta\varphi = \pm 2°$, $\Delta m = 2$ GeV. Thus the spectrometer enables us to map the e^+e^- mass region from 1 to 5 GeV in three overlapping settings.

Figure 1(b) shows the time-of-flight spectrum between the e^+ and e^- arms in the mass region $2.5 < m < 3.5$ GeV. A clear peak of 1.5-nsec width is observed. This enables us to reject the accidentals easily. Track reconstruction between the two arms was made and again we have a clear-cut distinction between real pairs and accidentals. Figure 1(c) shows the shower and lead-glass pulse height spectrum for the events in the mass region $3.0 < m < 3.2$ GeV. They are again in agreement with the calibration made by the e beam.

Typical data are shown in Fig. 2. There is a clear sharp enhancement at $m = 3.1$ GeV. Without folding in the 10^5 mapped magnetic points and the radiative corrections, we estimate a mass resolution of 20 MeV. As seen from Fig. 2 the width of the particle is consistent with zero.

To ensure that the observed peak is indeed a real particle ($J \to e^+e^-$) many experimental checks were made. We list seven examples:

(1) When we decreased the magnet currents by 10%, the peak remained fixed at 3.1 GeV (see Fig. 2).

(2) To check second-order effects on the target, we increased the target thickness by a factor of 2. The yield increased by a factor of 2, not by 4.

(3) To check the pileup in the lead glass and shower counters, different runs with different voltage settings on the counters were made. No effect was observed on the yield of J.

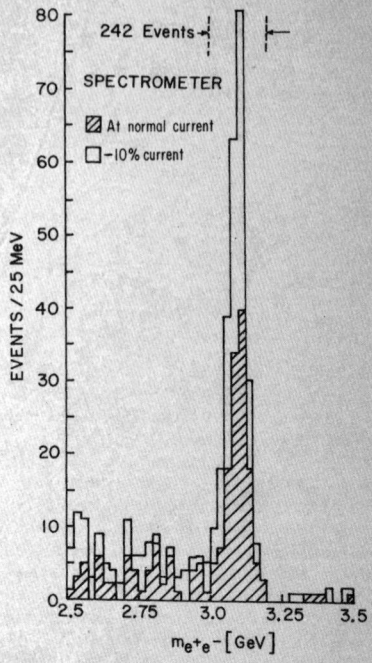

FIG. 2. Mass spectrum showing the existence of J. Results from two spectrometer settings are plotted showing that the peak is independent of spectrometer currents. The run at reduced current was taken two months later than the normal run.

(4) To ensure that the peak is not due to scattering from the sides of magnets, cuts were made in the data to reduce the effective aperture. No significant reduction in the J yield was found.

(5) To check the read-out system of the chambers and the triggering system of the hodoscopes, runs were made with a few planes of chambers deleted and with sections of the hodoscopes omitted from the trigger. No effect was observed on the J yield.

(6) Runs with different beam intensity were made and the yield did not change.

(7) To avoid systematic errors, half of the data were taken at each spectrometer polarity.

These and many other checks convinced us that we have observed a real massive particle $J \to ee$.

If we assume a production mechanism for J to be $d\sigma/dp_\perp \propto \exp(-6p_\perp)$ we obtain a yield of J of approximately 10^{-34} cm^2.

The most striking feature of J is the possibility that it may be one of the theoretically suggested charmed particles[2] or a's[3] or Z_0's,[4] etc. In order to study the real nature of J,[5] measurements are now underway on the various decay modes, e.g., an $e\pi\nu$ mode would imply that J is weakly interacting in nature.

It is also important to note the absence of an e^+e^- continuum, which contradicts the predictions of parton models.[6]

We wish to thank Dr. R. R. Rau and the alternating-gradient synchrotron staff who have done an outstanding job in setting up and maintaining this experiment. We thank especially Dr. F. Eppling, B. M. Bailey, and the staff of the Laboratory for Nuclear Science for their help and encouragement. We thank also Ms. I. Schulz, Ms. H. Feind, N. Feind, D. Osborne, G. Krey, J. Donahue, and E. D. Weiner for help and assistance. We thank also M. Deutsch, V. F. Weisskopf, T. T. Wu, S. Drell, and S. Glashow for many interesting conversations.

---

†Accepted without review under policy announced in Editorial of 20 July 1964 [Phys. Rev. Lett. 13, 79 (1964)].

[1]The first work on $p+p \to \mu^+ +\mu^- +x$ was done by L. M. Lederman et al., Phys. Rev. Lett. 25, 1523 (1970).
[2]S. L. Glashow, private communication.
[3]T. D. Lee, Phys. Rev. Lett. 26, 801 (1971).
[4]S. Weinberg, Phys. Rev. Lett. 19, 1264 (1967), and 27, 1688 (1971), and Phys. Rev. D 5, 1412, 1962 (1972).
[5]After completion of this paper, we learned of a similar result from SPEAR. B. Richter and W. Panofsky, private communication; J.-E. Augustin et al., following Letter [Phys. Rev. Lett. 33, 1404 (1974)].
[6]S. D. Drell and T. M. Yan, Phys. Rev. Lett. 25, 316 (1970). An improved version of the theory is not in contradiction with the data.

논문 웹페이지

위대한 논문과의 만남을 마무리하며

1960년대 초, 입자물리학은 혼란 그 자체였습니다. 가속기에서 쏟아지는 새로운 입자들은 마치 자연이 일부러 혼란을 조장하는 듯 무질서했고, 물리학자들은 이 수많은 입자를 어떻게 이해해야 할지 방향을 잃고 있었습니다. 어떤 이는 이를 '입자 동물원'이라고 불렀습니다.

그때 한 사람의 눈에만 이 혼란 속에서 질서가 보였습니다. 머리 겔만(Murray Gell-Mann). 그는 숫자와 패턴, 대칭을 사랑한 이론물리학자였습니다. 1964년 그는 세상을 놀라게 할 짧은 논문 한 편을 발표했습니다. 논문에서 그는 주장했습니다.

"이 모든 복잡한 입자들은, 사실 더 작은 단위인 쿼크(quark) 세 개로 이루어져 있을 수 있다."

그는 쿼크에 업(up), 다운(down), 스트레인지(strange)라는 이름도 붙였습니다. 낯설고도 낭만적인 이 명명은, 이론물리학이 시적 상상력과 수학적 정밀성을 모두 가지는 것을 보여주는 상징이 되었습니다. 수십 년이 지난 뒤 우리는 실험으로 겔만의 예측이 옳았음을 확인했습니다. 오늘날 표준모형(Standard Model)이라는 현대 입자물리학의 토대 위에는 그가 남긴 '쿼크모형'이 뼈대를 이룹니다.

이 책은 쿼크모형의 역사, 수학적 구조, 철학적 의미, 그리고 겔만 교수의 삶을 따라가는 여정입니다. 쿼크는 단순한 물리학 이론이 아닙니다. 보이지 않는 것을 믿는 인간 이성의 힘, 그 믿음을 수식과 실험으로 증명해 내는 과학의 승리를 상징합니다. 과연 보이지 않는 입자들로 이루어진 우리 세계는 무엇으로부터 시작된 것일까요? 이 책은 겔만이 꾼 그 상상의 꿈을 소개합니다.

이 시리즈의 출판 기획상 수식을 피할 수 없을 때는 고등학교 수학 정도를 아는 사람이라면 이해할 수 있도록 처음 쓴 원고를 고치고 또 고치는 작업을 반복했습니다. 그렇게 하여 수식을 줄여보려고 했습니다. 하지만 물리를 좋아하는 사람들이 쉽게 따라갈 수 있도록 친절하게 설명했습니다.

원고를 쓰기 위해 20세기의 여러 논문을 뒤적거렸습니다. 지금과는 완연히 다른 용어와 기호 때문에 많이 힘들었습니다. 특히 번역이 안 되어 있는 자료들이 많았지만 프랑스 논문에 대해서는 불문과를 졸업한 아내의 도움으로 조금은 이해할 수 있었습니다.

드디어 〈노벨상 수상자들의 오리지널 논문으로 배우는 과학〉 시리즈 20권이 완성되었습니다. 20권이라는 여정의 끝에서 다시 처음의 질문으로 돌아갑니다.

"과학이란 무엇인가?" "왜 보이지 않는 세계를 실명하려 하는가?"

이 시리즈에는 과학자들의 직관, 의심, 상상력, 그리고 진실을 향한 고집스러운 집념이 담겨 있습니다.

제가 느끼는 과학에 대한 즐거움을 독자들이 공유할 수 있기를 바라며 이제 힘들었지만 재미있었던 〈노벨상 수상자들의 오리지널 논문으로 배우는 과학〉 시리즈와의 씨름을 여기서 멈추려고 합니다.

이제 마지막 책을 마무리하며 독자 여러분에게 하고 싶은 말이 있습니다.
"이 여정에 함께해 주셔서 고맙습니다."

진주에서 정완상 교수

이 책을 위해 참고한 논문들

1장

[1] J. Cockcroft and E. Walton, "Disintegration of Lithium by Swift Protons", Nature. 129 (649): 649, April 1932.

[2] G. Ising, "Prinzip einer Methode zur Herstellung von Kanalstrahlen hoher Voltzahl", Arkiv för matematik. astronomi och fysik (in German). 18 (30): 1-4, 1924.

[3] R. Wideröe, "Über ein neues Prinzip zur Herstellung hoher Spannungen", Archiv für Elektrotechnik (in German). 21 (4): 387-406, 1928.

[4] E. O. Lawrence and M. S. Livingston, "The Production of High Speed Light Ions Without the Use of High Voltages", Physical Review. 40 (1). American Physical Society: 19-35, 1 April 1932.

2장

[1] É. Galois, "Mémoire sur les conditions de résolubilité des équations par radicaux", Journal de mathématiques pures et appliquées. Tome XI. 417–433, 1846.

3장

[1] J. Thomson, "Cathode Rays", Phil. Mag. 5; 293, 1897.

[2] W. Heisenberg, "Über den Bau der Atomkerne", Zeitschrift für Physik (in German). 77 (1-2): 1-11, 1932.

[3] H. Yukawa, "On the Interaction of Elementary Particles", Proc. Phys.-Math. Soc. Jpn. 17 (48), 1935.

[4] C. M. G. Lattes; H. Muirhead; G. P. S. Occhialini; C. F. Powell, "Processes Involving Charged Mesons", Nature. 159 (4047): 694, 1947.

4장

[1] G. D. Rochester and C. C. Butler, "Evidence for the Existence of New Unstable Elementary Particles", Nature. 160 (4077): 855-857, 1947.

[2] T. D. Lee and C. N. Yang, "Question of Parity Conservation in Weak Interactions", Physical Review. 104 (1): 254, 1 October 1956.

[3] T. Nakano and N. Nishijima, "Charge Independence for V-particles", Progress of Theoretical Physics. 10 (5): 581, 1953.

[4] M. Gell-Mann, Phys. Rev. 92; 833, 1953.

[5] M. Gell-Mann, "The Interpretation of the New Particles as

Displaced Charged Multiplets", Il Nuovo Cimento. 4 (S2): 848-866, 1956.

5장

[1] S. Sakata, On a Composite Model for the New Particles, Progress of Theoretical Physics. Volume 16. Issue 6. 686-688, December 1956.

[2] E. Fermi and C. N. Yang, Phys. Rev. 76; 1739–1743, 1949.

[3] M. Gell−Mann, "A Schematic Model of Baryons and Mesons", Physics Letters. 8 (3): 214-215, 1964.

[4] G. Zweig, "An SU(3) Model for Strong Interaction Symmetry and its Breaking", CERN Document Server. CERN−TH−401, 1964.

[5] Barnes et al., "Observation of a Hyperon with Strangeness Number Three", Physical Review Letters. 12 (8): 204, 1964.

6장

[1] O. W. Greenberg, "Spin and unitary−spin independence in a paraquark model of baryons and mesons", Physical Review Letters. 13 (20): 598-602, 1964.

[2] M. Y. Han and Y. Nambu, "Three−triplet model with double SU(3) symmetry", Physical Review B. 139 (4B): 1006, 1965.

수식에 사용하는 그리스 문자

대문자	소문자	읽기	대문자	소문자	읽기
A	α	알파(alpha)	N	ν	뉴(nu)
B	β	베타(beta)	Ξ	ξ	크시(xi)
Γ	γ	감마(gamma)	O	o	오미크론(omicron)
Δ	δ	델타(delta)	Π	π	파이(pi)
E	ε	엡실론(epsilon)	P	ρ	로(rho)
Z	ζ	제타(zeta)	Σ	σ	시그마(sigma)
H	η	에타(eta)	T	τ	타우(tau)
Θ	θ	세타(theta)	Y	υ	입실론(upsilon)
I	ι	요타(iota)	Φ	φ	피(phi)
K	κ	카파(kappa)	X	χ	키(chi)
Λ	λ	람다(lambda)	Ψ	ψ	프시(psi)
M	μ	뮤(mu)	Ω	ω	오메가(omega)

노벨 물리학상 수상자들을 소개합니다

이 책에 언급된 노벨상 수상자는 이름 앞에 ★로 표시하였습니다.

연도	수상자	수상 이유
1901	빌헬름 콘라트 뢴트겐	그의 이름을 딴 놀라운 광선의 발견으로 그가 제공한 특별한 공헌을 인정하여
1902	헨드릭 안톤 로런츠	복사 현상에 대한 자기의 영향에 대한 연구를 통해 그들이 제공한 탁월한 공헌을 인정하여
	피터르 제이만	
1903	앙투안 앙리 베크렐	자발 방사능 발견으로 그가 제공한 탁월한 공로를 인정하여
	피에르 퀴리	앙리 베크렐 교수가 발견한 방사선 현상에 대한 공동 연구를 통해 그들이 제공한 탁월한 공헌을 인정하여
	★마리 퀴리	
1904	존 윌리엄 스트럿 레일리	가장 중요한 기체의 밀도에 대한 조사와 이러한 연구와 관련하여 아르곤을 발견한 공로
1905	필리프 레나르트	음극선에 대한 연구
1906	★조지프 존 톰슨	기체에 의한 전기 전도에 대한 이론적이고 실험적인 연구의 큰 장점을 인정하여
1907	앨버트 에이브러햄 마이컬슨	광학 정밀 기기와 그 도움으로 수행된 분광 및 도량형 조사
1908	가브리엘 리프만	간섭 현상을 기반으로 사진적으로 색상을 재현하는 방법
1909	굴리엘모 마르코니	무선 전신 발전에 기여한 공로를 인정받아
	카를 페르디난트 브라운	
1910	요하네스 디데릭 판데르발스	기체와 액체의 상태 방정식에 관한 연구
1911	빌헬름 빈	열복사 법칙에 관한 발견
1912	닐스 구스타프 달렌	등대와 부표를 밝히기 위해 가스 어큐뮬레이터와 함께 사용하기 위한 자동 조절기 발명

노벨 물리학상 수상자 목록

1913	헤이커 카메를링 오너스	특히 액체 헬륨 생산으로 이어진 저온에서의 물질 특성에 대한 연구
1914	막스 폰 라우에	결정에 의한 X선 회절 발견
1915	윌리엄 헨리 브래그 윌리엄 로런스 브래그	X선을 이용한 결정구조 분석에 기여한 공로
1916	수상자 없음	
1917	찰스 글러버 바클라	원소의 특징적인 뢴트겐 복사 발견
1918	막스 플랑크	에너지 양자 발견으로 물리학 발전에 기여한 공로 인정
1919	요하네스 슈타르크	커낼선의 도플러 효과와 전기장에서 분광선의 분할 발견
1920	샤를 에두아르 기욤	니켈강 합금의 이상 현상을 발견하여 물리학의 정밀 측정에 기여한 공로를 인정하여
1921	알베르트 아인슈타인	이론 물리학에 대한 공로, 특히 광전효과 법칙 발견
1922	닐스 보어	원자 구조와 원자에서 방출되는 방사선 연구에 기여
1923	로버트 앤드루스 밀리컨	전기의 기본 전하와 광전효과에 관한 연구
1924	칼 만네 예오리 시그반	X선 분광학 분야에서의 발견과 연구
1925	제임스 프랑크 구스타프 헤르츠	전자가 원자에 미치는 영향을 지배하는 법칙 발견
1926	장 바티스트 페랭	물질의 불연속 구조에 관한 연구, 특히 침전 평형 발견
1927	아서 콤프턴	그의 이름을 딴 효과 발견
	★찰스 톰슨 리스 윌슨	수증기 응축을 통해 전하를 띤 입자의 경로를 볼 수 있게 만든 방법
1928	오언 윌런스 리처드슨	열전자 현상에 관한 연구, 특히 그의 이름을 딴 법칙 발견
1929	루이 드브로이	전자의 파동성 발견
1930	찬드라세카라 벵카타 라만	빛의 산란에 관한 연구와 그의 이름을 딴 효과 발견
1931	수상자 없음	

1932	★베르너 하이젠베르크	수소의 동소체 형태 발견으로 이어진 양자역학의 창시
1933	에르빈 슈뢰딩거	원자 이론의 새로운 생산적 형태 발견
	★폴 디랙	
1934	수상자 없음	
1935	★제임스 채드윅	중성자 발견
1936	빅토르 프란츠 헤스	우주 방사선 발견
	★칼 데이비드 앤더슨	양전자 발견
1937	클린턴 조지프 데이비슨	결정에 의한 전자의 회절에 대한 실험적 발견
	조지 패짓 톰슨	
1938	★엔리코 페르미	중성자 조사에 의해 생성된 새로운 방사성 원소의 존재에 대한 시연 및 이와 관련된 느린중성자에 의한 핵반응 발견
1939	★어니스트 로런스	사이클로트론의 발명과 개발, 특히 인공 방사성 원소와 관련하여 얻은 결과
1940	수상자 없음	
1941		
1942		
1943	오토 슈테른	분자선 방법 개발 및 양성자의 자기 모멘트 발견에 기여
1944	★이지도어 아이작 라비	원자핵의 자기적 특성을 기록하기 위한 공명 방법
1945	★볼프강 파울리	파울리 원리라고도 불리는 배제 원리의 발견
1946	퍼시 윌리엄스 브리지먼	초고압을 발생시키는 장치의 발명과 고압 물리학 분야에서 그가 이룬 발견에 대해
1947	에드워드 빅터 애플턴	대기권 상층부의 물리학 연구, 특히 이른바 애플턴층의 발견
1948	패트릭 메이너드 스튜어트 블래킷	윌슨 구름상자 방법의 개발과 핵물리학 및 우주 방사선 분야에서의 발견
1949	★유카와 히데키	핵력에 관한 이론적 연구를 바탕으로 중간자 존재 예측

연도	수상자	업적
1950	★세실 프랭크 파월	핵 과정을 연구하는 사진 방법의 개발과 이 방법으로 만들어진 중간자에 관한 발견
1951	★존 더글러스 콕크로프트 ★어니스트 토머스 신턴 월턴	인위적으로 가속된 원자 입자에 의한 원자핵 변환에 대한 선구자적 연구
1952	펠릭스 블로흐 에드워드 밀스 퍼셀	핵자기 정밀 측정을 위한 새로운 방법 개발 및 이와 관련된 발견
1953	프리츠 제르니커	위상차 방법 시연, 특히 위상차 현미경 발명
1954	막스 보른	양자역학의 기초 연구, 특히 파동함수의 통계적 해석
	발터 보테	우연의 일치 방법과 그 방법으로 이루어진 그의 발견
1955	윌리스 유진 램	수소 스펙트럼의 미세 구조에 관한 발견
	폴리카프 쿠시	전자의 자기 모멘트를 정밀하게 측정한 공로
1956	윌리엄 브래드퍼드 쇼클리 존 바딘 월터 하우저 브래튼	반도체 연구 및 트랜지스터 효과 발견
1957	★양전닝 ★리정다오	소립자에 관한 중요한 발견으로 이어진 소위 패리티 법칙에 대한 철저한 조사
1958	파벨 알렉세예비치 체렌코프 일리야 프랑크 이고리 탐	체렌코프 효과의 발견과 해석
1959	★에밀리오 지노 세그레 ★오언 체임벌린	반양성자 발견
1960	도널드 아서 글레이저	거품 상자의 발명
1961	로버트 호프스태터	원자핵의 전자 산란에 대한 선구적인 연구와 핵자 구조에 관한 발견
	루돌프 뫼스바워	감마선의 공명 흡수에 관한 연구와 그의 이름을 딴 효과에 대한 발견

1962	레프 다비도비치 란다우	응집 물질, 특히 액체 헬륨에 대한 선구적인 이론
1963	★유진 폴 위그너	원자핵 및 소립자 이론에 대한 공헌, 특히 기본 대칭 원리의 발견 및 적용을 통한 공로
	마리아 괴페르트 메이어	핵 껍질 구조에 관한 발견
	한스 옌센	
1964	니콜라이 바소프	메이저-레이저 원리에 기반한 발진기 및 증폭기의 구성으로 이어진 양자 전자 분야의 기초 작업
	알렉산드르 프로호로프	
	찰스 하드 타운스	
1965	★도모나가 신이치로	소립자 물리학에 심층적인 결과를 가져온 양자전기역학의 근본적인 연구
	★줄리언 슈윙거	
	★리처드 필립스 파인먼	
1966	알프레드 카스틀러	원자에서 헤르츠 공명을 연구하기 위한 광학적 방법의 발견 및 개발
1967	한스 알브레히트 베테	핵반응 이론, 특히 별의 에너지 생산에 관한 발견에 기여
1968	★루이스 월터 앨버레즈	소립자 물리학에 대한 결정적인 공헌, 특히 수소 기포 챔버 사용 기술 개발과 데이터 분석을 통해 가능해진 다수의 공명 상태 발견
1969	★머리 겔만	기본 입자의 분류와 그 상호 작용에 관한 공헌 및 발견
1970	한네스 올로프 예스타 알벤	플라스마 물리학의 다양한 부분에서 유익한 응용을 통해 자기유체역학의 기초 연구 및 발견
	루이 외젠 펠릭스 네엘	고체물리학에서 중요한 응용을 이끈 반강자성 및 강자성에 관한 기초 연구 및 발견
1971	데니스 가보르	홀로그램 방법의 발명 및 개발
1972	존 바딘	일반적으로 BCS 이론이라고 하는 초전도 이론을 공동으로 개발한 공로
	리언 닐 쿠퍼	
	존 로버트 슈리퍼	

연도	수상자	업적
1973	에사키 레오나	반도체와 초전도체의 터널링 현상에 관한 실험적 발견
	이바르 예베르	
	브라이언 데이비드 조지프슨	터널 장벽을 통과하는 초전류 특성, 특히 일반적으로 조지프슨 효과로 알려진 현상에 대한 이론적 예측
1974	마틴 라일	전파 천체물리학의 선구적인 연구: 라일은 특히 개구 합성 기술의 관찰과 발명, 그리고 휴이시는 펄서 발견에 결정적인 역할을 함
	앤터니 휴이시	
1975	오게 닐스 보어	원자핵에서 집단 운동과 입자 운동 사이의 연관성 발견과 이 연관성에 기초한 원자핵 구조 이론 개발
	벤 로위 모텔손	
	제임스 레인워터	
1976	★버턴 릭터	새로운 종류의 무거운 기본 입자 발견에 대한 선구적인 작업
	★새뮤얼 차오 충 팅	
1977	필립 워런 앤더슨	자기 및 무질서 시스템의 전자 구조에 대한 근본적인 이론적 조사
	네빌 프랜시스 모트	
	존 해즈브룩 밴블렉	
1978	표트르 레오니도비치 카피차	저온 물리학 분야의 기본 발명 및 발견
	아노 앨런 펜지어스	우주 마이크로파 배경 복사의 발견
	로버트 우드로 윌슨	
1979	★셸던 리 글래쇼	특히 약한 중성 전류의 예측을 포함하여 기본 입자 사이의 통일된 약한 전자기 상호 작용 이론에 대한 공헌
	★압두스 살람	
	★스티븐 와인버그	
1980	제임스 왓슨 크로닌	중성 K 중간자의 붕괴에서 기본 대칭 원리 위반 발견
	밸 로그즈던 피치	
1981	니콜라스 블룸베르헌	레이저 분광기 개발에 기여
	아서 레너드 숄로	
	카이 만네 뵈리에 시그반	고해상도 전자 분광기 개발에 기여

연도	수상자	업적
1982	케네스 게디스 윌슨	상전이와 관련된 임계 현상에 대한 이론
1983	수브라마니안 찬드라세카르	별의 구조와 진화에 중요한 물리적 과정에 대한 이론적 연구
	윌리엄 앨프리드 파울러	우주의 화학 원소 형성에 중요한 핵반응에 대한 이론 및 실험적 연구
1984	★카를로 루비아	약한 상호 작용의 커뮤니케이터인 필드 입자 W와 Z의 발견으로 이어진 대규모 프로젝트에 결정적인 기여
	시몬 판데르 메이르	
1985	클라우스 폰 클리칭	양자화된 홀 효과의 발견
1986	에른스트 루스카	전자 광학의 기초 작업과 최초의 전자 현미경 설계
	게르트 비니히	스캐닝 터널링 현미경 설계
	하인리히 로러	
1987	요하네스 게오르크 베드노르츠	세라믹 재료의 초전도성 발견에서 중요한 돌파구
	카를 알렉산더 뮐러	
1988	★리언 레더먼	뉴트리노 빔 방법과 뮤온 중성미자 발견을 통한 경입자의 이중 구조 증명
	★멜빈 슈워츠	
	★잭 스타인버거	
1989	노먼 포스터 램지	분리된 진동 필드 방법의 발명과 수소 메이저 및 기타 원자시계에서의 사용
	한스 게오르크 데멜트	이온 트랩 기술 개발
	볼프강 파울	
1990	제롬 프리드먼	입자 물리학에서 쿼크 모델 개발에 매우 중요한 역할을 한 양성자 및 구속된 중성자에 대한 전자의 심층 비탄성 산란에 관한 선구적인 연구
	헨리 웨이 켄들	
	리처드 테일러	
1991	피에르질 드 젠	간단한 시스템에서 질서 현상을 연구하기 위해 개발된 방법을 보다 복잡한 형태의 물질, 특히 액정과 고분자로 일반화할 수 있음을 발견

노벨 물리학상 수상자 목록

연도	수상자	업적
1992	조르주 샤르파크	입자 탐지기, 특히 다중 와이어 비례 챔버의 발명 및 개발
1993	러셀 헐스 조지프 테일러	새로운 유형의 펄서 발견, 중력 연구의 새로운 가능성을 연 발견
1994	버트럼 브록하우스	중성자 분광기 개발
	클리퍼드 셜	중성자 회절 기술 개발
1995	★마틴 펄	타우 렙톤의 발견
	★프레더릭 라이너스	중성미자 검출
1996	데이비드 리 더글러스 오셔로프 로버트 리처드슨	헬륨-3의 초유동성 발견
1997	스티븐 추 클로드 코엔타누지 윌리엄 필립스	레이저 광으로 원자를 냉각하고 가두는 방법 개발
1998	로버트 로플린 호르스트 슈퇴르머 대니얼 추이	부분적으로 전하를 띤 새로운 형태의 양자 유체 발견
1999	헤라르뒤스 엇호프트 마르티뉴스 펠트만	물리학에서 전기약력 상호작용의 양자 구조 규명
2000	조레스 알표로프 허버트 크로머	정보 통신 기술에 대한 기초 작업(고속 및 광전자 공학에 사용되는 반도체 이종 구조 개발)
	잭 킬비	정보 통신 기술에 대한 기초 작업(집적회로 발명에 기여)
2001	에릭 코넬 칼 위먼 볼프강 케테를레	알칼리 원자의 희석 가스에서 보스-아인슈타인 응축 달성 및 응축 특성에 대한 초기 기초 연구

연도	수상자	업적
2002	레이먼드 데이비스	천체물리학, 특히 우주 중성미자 검출에 대한 선구적인 공헌
	고시바 마사토시	
	리카르도 자코니	우주 X선 소스의 발견으로 이어진 천체물리학에 대한 선구적인 공헌
2003	알렉세이 아브리코소프	초전도체 및 초유체 이론에 대한 선구적인 공헌
	비탈리 긴즈부르크	
	앤서니 레깃	
2004	★데이비드 그로스	강한 상호작용 이론에서 점근적 자유의 발견
	★데이비드 폴리처	
	★프랭크 윌첵	
2005	로이 글라우버	광학 일관성의 양자 이론에 기여
	존 홀	광 주파수 콤 기술을 포함한 레이저 기반 정밀 분광기 개발에 기여
	테오도어 헨슈	
2006	존 매더	우주 마이크로파 배경 복사의 흑체 형태와 이방성 발견
	조지 스무트	
2007	알베르 페르	자이언트 자기 저항의 발견
	페터 그륀베르크	
2008	★난부 요이치로	아원자 물리학에서 자발적인 대칭 깨짐 메커니즘 발견
	★고바야시 마코토	자연계에 적어도 세 종류의 쿼크가 존재함을 예측한 깨진 대칭의 기원 발견
	★마스카와 도시히데	
2009	찰스 가오	광 통신을 위한 섬유의 빛 전송에 관한 획기적인 업적
	윌러드 보일	영상 반도체 회로(CCD 센서)의 발명
	조지 엘우드 스미스	
2010	안드레 가임	2차원 물질 그래핀에 관한 획기적인 실험
	콘스탄틴 노보셀로프	

연도	수상자	업적
2011	솔 펄머터 브라이언 슈밋 애덤 리스	원거리 초신성 관측을 통한 우주 가속 팽창 발견
2012	세르주 아로슈 데이비드 와인랜드	개별 양자 시스템의 측정 및 조작을 가능하게 하는 획기적인 실험 방법
2013	프랑수아 앙글레르 ★피터 힉스	아원자 입자의 질량 기원에 대한 이해에 기여하고 최근 CERN의 대형 하드론 충돌기에서 ATLAS 및 CMS 실험을 통해 예측된 기본 입자의 발견을 통해 확인된 메커니즘의 이론적 발견
2014	아카사키 이사무 아마노 히로시 나카무라 슈지	밝고 에너지 절약형 백색 광원을 가능하게 한 효율적인 청색 발광 다이오드의 발명
2015	가지타 다카아키 아서 맥도널드	중성미자가 질량을 가지고 있음을 보여주는 중성미자 진동 발견
2016	데이비드 사울레스 덩컨 홀데인 마이클 코스털리츠	위상학적 상전이와 물질의 위상학적 위상에 대한 이론적 발견
2017	라이너 바이스 킵 손 배리 배리시	LIGO 탐지기와 중력파 관찰에 결정적인 기여
2018	아서 애슈킨 제라르 무루 도나 스트리클런드	레이저 물리학 분야의 획기적인 발명(광학 핀셋과 생물학적 시스템에 내한 응용) 레이저 물리학 분야의 획기적인 발명(고강도 초단파 광 펄스 생성 방법)
2019	제임스 피블스 미셸 마요르 디디에 쿠엘로	우주의 진화와 우주에서 지구의 위치에 대한 이해에 기여(물리 우주론의 이론적 발견) 우주의 진화와 우주에서 지구의 위치에 대한 이해에 기여(태양형 항성 주위를 공전하는 외계 행성 발견)

연도	수상자	업적
2020	로저 펜로즈	블랙홀 형성이 일반 상대성 이론의 확고한 예측이라는 발견
	라인하르트 겐첼	우리 은하의 중심에 있는 초거대 밀도 물체 발견
	앤드리아 게즈	
2021	마나베 슈쿠로	복잡한 시스템에 대한 이해에 획기적인 기여(지구 기후의 물리적 모델링, 가변성을 정량화하고 지구 온난화를 안정적으로 예측)
	클라우스 하셀만	
	조르조 파리시	복잡한 시스템에 대한 이해에 획기적인 기여 (원자에서 행성 규모에 이르는 물리적 시스템의 무질서와 요동의 상호작용 발견)
2022	알랭 아스페	얽힌 광자를 사용한 실험, 벨 불평등 위반 규명 및 양자 정보 과학 개척
	존 클라우저	
	안톤 차일링거	
2023	피에르 아고스티니	물질의 전자 역학 연구를 위해 아토초(100경분의 1초) 빛 펄스를 생성하는 실험 방법 고안
	페렌츠 크라우스	
	안 륄리에	
2024	존 홉필드	인공신경망을 이용해 머신러닝을 가능하게 하는 기초적인 발견과 발명
	제프리 힌턴